A Glossary of Corrosion-Related Terms
Used in Science and Industry

Mark S. Vukasovich
Editor

Published by:
Society of Automotive Engineers, Inc.
400 Commonwealth Drive
Warrendale, PA 15096-0001
U.S.A.
Phone: (412) 776-4841
Fax: (412) 776-5760

Library of Congress Cataloging-in-Publication Data

A Glossary of corrosion-related terms used in science and industry /
  Mark S. Vukasovich, editor.
    p.   cm.
    ISBN 1-56091-590-0
    1. Corrosion and anti-corrosives—Dictionaries.   I. Vukasovich, Mark S.
  TA418.74.G58   1995
  620.1'1223—dc20   94-43488
                                                                                      CIP

Copyright ©1995 Society of Automotive Engineers, Inc.

ISBN 1-56091-590-0

All rights reserved. Printed in the United States of America.

Permission to photocopy for internal or personal use, or the internal or personal use of specific clients, is granted by SAE for libraries and other users registered with the Copyright Clearance Center (CCC), provided that the base fee of $.50 per page is paid directly to CCC, 222 Rosewood Dr., Danvers, MA 01923. Special requests should be addressed to the SAE Publications Group. 1-56091-590-0 /95 $.50.

SAE Order No. R-152

Preface

A Glossary of Corrosion-Related Terms Used in Science and Industry was compiled to provide those studying or confronting corrosion with definitions and descriptions of the terms they might utilize or encounter in their work. It is intended to be a desk reference for students, scientists, engineers, and laymen needing ready access to these idioms. The glossary is comprehensive. More than 4000 terms cover scientific and empirical expressions, trade words, and lesser-known jargon unique to various industries and materials subject to corrosion. The glossary also includes terms from various engineering disciplines, chemistry, electrochemistry, physics, metallurgy, biology, and earth sciences which are used in discussing corrosion, its processes, appearances, causes, controls, testing, evaluation, and environmental impact. Synonyms, acronyms, and abbreviations in the corrosion literature are also covered. Proprietary names are restricted to a few in long-term or widespread usage.

This glossary began years ago as a personal compilation collected as I came upon them in a wide range of dictionaries, encyclopedias, word lists, books, journal articles, trade publications, and in government and industry documents relating to methods, standards and specifications. I undertook the compilation to use in my own industrial research and teaching careers since no such comprehensive glossary was then printed. Believing that others might benefit from the collection, I decided subsequently to expand the effort and put together this glossary for publication. Except for its fundamental scientific expressions, the definitions and descriptions reflect my best interpretation of a consensus of their published phraseology, specificity, and expansiveness so as to be useful to a wide audience. While believed accurate, the terms are not represented as standards or as their only correct pronouncement by either me or the publisher. Both SAE and I welcome suggestions for improvement of this work.

Mark S. Vukasovich
August 1994

A

AA—abbreviation for the Aluminum Association.

AAR—abbreviation for the Association of American Railroads.

AAS—abbreviation for atomic absorption spectroscopy; an analytical technique in which a sample is vaporized and the nonexcited atoms absorb electromagnetic radiation at characterizing wavelengths.

abatement—reduction in degree or intensity of pollution.

ablative antifouling coating—antifouling coating that slowly dissolves in seawater to expose new layers of the coating.

abrasion—process of wearing down or rubbing away by means of friction. *See also* **wear**.

abrasion corrosion—conjoint action involving abrasion and corrosion.

abrasion resistance—ability of a material to resist being worn away and to maintain its original appearance and structure when subjected to rubbing, scraping, or impingement.

abrasive—material used in abrasive blast cleaning, e.g., sand, grit, glass beads, steel shot, etc.

abrasive blast cleaning—process for cleaning or finishing a surface by which an abrasive material is directed at high velocity against the workpiece.

abrasive wear—mechanical degradation of a surface of a material due to its relative motion against another surface or from contact with solid particles.

ABS—abbreviation for an alkyl benzene sulfonate basis detergent. *See* **alkyl benzene sulfonate**.

ABS plastic—abbreviation for any of a class of plastics based on acrylonitrile-butadiene-styrene copolymers.

absolute electrode potential—hypothetical electrode potential value expressed with respect to a zero-potential reference electrode. Such a reference electrode does not exist, hence it is hypothetical.

absolute humidity—mass of water vapor present in a unit volume of air. Also known as specific humidity.

absolute temperature scale—primary temperature standard based on the temperature at which water, ice, and water vapor coexist in thermal equilibrium (the triple point of water) and which condition exists only at a pressure of 4.58 torr. The corresponding temperature is arbitrarily defined to be 273.16 kelvin. The magnitude of the kelvin unit is the same as the Celsius degree (°C).

absolute zero—lower limit to the temperature scale. Defined as zero on the kelvin scale (0 K) and is equal to –273.15 °C or –459.67 °F. It is the temperature at which a gas would show no pressure if the general law for gases (ideal gas) would hold for all temperatures.

absorbate—material retained by the process of absorption.

absorbent—material in which absorption occurs.

absorption—take-up of a gas by a solid or liquid or the take-up of a liquid by a solid without chemical reaction and the process in which the absorbed substance permeates the bulk of the absorbing substance. *Compare with* **adsorption**.

absorption refrigeration—refrigeration system that mainly uses water as the refrigerant and lithium bromide as the absorbent or, for low-temperature service, water and ammonia, respectively. *See also* **refrigeration**.

absorption test—term used in concrete technology. One of several possible specified test methods for determining absorption by concrete.

AC (a.c.)—*abbreviation for* **alternating current**.

AC impedance (electrochemical impedance technique)—electrochemical technique applicable to corroding systems for corrosion prediction, control, and mechanism studies. AC impedance measurements allow for separation and independent analysis of the resistive and capacitive elements of the electrochemical corrosion reaction.

accelerated aging—deterioration of a material faster than normal because of being subjected to artificial conditions emphasizing one or more of the aspects of normal aging and which aspects are specified by a test method.

accelerated corrosion test—method designed to simulate, in a shorter time, the deteriorating effect of normal, longer-term service.

accelerated life test—method designed to estimate a service life for a material in a shorter time frame than its normal usage would be by subjecting it to an artificial exaggeration of one or more of the deteriorating effects of its normal, long-term service conditions.

accelerated weathering—test method designed to simulate, in shorter time, the deteriorating action of natural outdoor weathering on coatings by intensifying certain exposure parameters.

acceleration period—relates to cavitation and liquid impingement erosion. It is the stage following the incubation period during which the erosion rate increases from near zero to a maximum value.

accelerator—substance that increases the rate of a chemical reaction or cure. Often used interchangeably with the terms promoter and catalyst.

acceptance testing—testing by the purchaser of received products to determine whether the quality of the manufactured products meets specific requirements.

accuracy—indication of the reliability of a measurement or observation. *See also under* **statistical terms**.

acetal—class of organic compounds that results when an aldehyde combines with two molecules of alcohol. The general formula for an acetal is $RCH(OR')_2$.

acetal resins—linear polymers of an aldehyde, commonly formaldehyde or butyraldehyde, in which an oxygen atom joins the repeating units in an ether rather than ester-type link.

acetals—organic compounds formed by addition of aldehyde molecules to alcohol molecules. If one molecule of aldehyde reacts with one molecule of alcohol, a hemiacetal is formed ($RCH(OH)OR'$). Further reaction with a second alcohol molecule produces a full acetal ($RCH(OR')_2$).

acetic acid salt spray (fog) test—accelerated corrosion test method applicable to ferrous and nonferrous metals with inorganic and organic coatings. See ASTM B 287 for details.

ACGIH—abbreviation for the American Conference of Governmental Industrial Hygienists.

ACI—abbreviation for the American Concrete Institute.

acid—(1) compound that yields hydrogen ions (H^+) when dissolved in water (Arrhenius theory). (2) compound or ion that is a proton donor (Lowery-Bronsted theory). (3) compound or atom that can accept a pair of electrons (Lewis theory). *Contrast with* **base**.

acid anhydride—nonmetal oxide compound that reacts with water to form an acid, e.g., carbon dioxide (CO_2), to form carbonic acid (H_2CO_3).

acid-base indicator—dye that changes color in response to the pH of a solution.

acid brick—brick made from fireclay that is a heat-resistant clay having a higher silica content than ordinary firebrick. Used to line vessels to impart corrosion resistance against hot acid or erosion-corrosion attack.

acid cleaning—process for removing hardness scales and metallic oxide corrosion products from heat exchangers by contacting with aqueous solutions of such mineral acids as hydrochloric, sulfuric, sulfamic, phosphoric, or nitric, or with their acid salts, or with organic acids such as citric, hydroxyacetic, or formic. Also process using a solution of a mineral acid, organic acid, or acid salt in combination with a wetting agent and detergent to remove oxide, shop soil, oil, grease, and other contaminants from metal surfaces, with or without the application of heat, before plating, painting, or storage. Differs from acid pickling in matter of degree.

acid embrittlement—hydrogen embrittlement when induced by acid cleaning treatment. *See also* **hydrogen embrittlement**.

acid etch—to clean or alter a surface by treatment with a suitable acid.

acid feed—cooling water treatment using acid, usually sulfuric acid, for the control of calcium carbonate fouling. Also the addition of an acid to obtain the most effective use of chlorine for disinfection.

acid foaming—procedure to clean heat exchangers (removed from service) by employing a foam generator to produce a highly acidic foam that is pumped through the heat-exchanger tubes.

acid fume resistance test—laboratory test for coated metal panels wherein they are placed in a closed chamber containing humidified acid fumes, at room temperature, for a specified time. Hydrochloric, sulfuric, sulfurous, or acetic acids are most frequently used for the corrosive fumes.

acid groups—functional groups having the properties of acids and most commonly applied to carboxyl groups (–COOH).

acidic—term most commonly used to describe a compound or solution having an excess of hydrogen ions.

acidified synthetic sea water test—see ASTM G 85 for details.

acid inhibitors—corrosion inhibitors for systems in which acids are likely to attack metal, e.g., when using acids to remove scale resulting from hot fabrication and for descaling water pipes. The inhibitors are mainly organic molecules containing nitrogen and sulfur in polar groups that adsorb on the bare metal surface to shield it from further acid attack. Common acid inhibitors include unsubstituted and substituted quinolines and thioureas.

acidity—quantitative capacity of aqueous media to react with hydroxyl ions.

acidity, free mineral—*see* **free mineral acidity**.

acid mine waters—acidic waters present in some underground coal mines as a result of the aerial and microbial oxidation of pyritic sulfur present in the coal seams or related strata. Their extreme corrosivity results primarily from their free acidity and presence of high concentrations of ferric and sulfate ions.

acid number (acid value)—numerical index of free acid in an oil, resin, varnish, or fat. It is equal to the number of milligrams of potassium hydroxide needed to neutralize one gram of the material.

acid rain—rain having a pH less than 5.6; usually the result of the reaction of atmospheric moisture with sulfur oxide gases from industrial emissions and with nitrogen oxide gases from car exhausts.

acid resistance—ability of a material to resist attack by acids, usually mineral acids.

acid salt—salt of a polybasic acid in which one or more of its hydrogen atoms has not been replaced by other cations.

acid slugging—flushing a vessel or conduit with an acidic solution for the purpose of dissolving scale while the vessel or conduit is still on-stream.

acid water—water having a pH value below 4.4 to 4.6, which is the end point of the methyl orange titration. Acid water is said to have a free mineral acidity.

acoustic emission testing (AE)—method for determining the structural integrity of a metal by analyzing the elastic waves generated by it when stressed. It is a nondestructive testing technique used for detecting leaks, cavitation, corrosion fatigue, pitting, and stress-corrosion cracking in vessels and lines.

acrolein—the organic compound $CH_2=CHCHO$. Also the name given to a gaseous class of low molecular weight organic compounds called alpha-, beta-unsaturated aldehydes. Used as microbiocides in cooling waters.

acrylate—ester or salt formed from acrylic acid ($CH_2=CHCOOH$).

acrylic latex—aqueous dispersions of either thermoplastic or thermosetting polymers or copolymers of acrylic acid, methacrylic acid, esters of these acids, or from acrylonitrile.

acrylic resin—family of synthetic resins, either thermoplastic or thermosetting, made by polymerizing esters or other derivatives of acrylic or methacrylic acid.

acrylonitrile—raw material of the formula $CH_2=CHCN$ used for the manufacture of synthetic resins and rubbers.

ACS—abbreviation for the American Chemical Society.

actinic radiation—electromagnetic radiation that is capable of initiating a chemical reaction.

actinide series—elements of atomic numbers 89 to 103 and analogous to the lanthanide series of the so-called rare earths.

activate—to put into a state of increased chemical activity.

activated alumina—aluminum oxide, made by heating hydrated aluminum hydroxide, having a variable porosity, particle size, and number of remaining –OH groups. One use for activated alumina is as a desiccant, absorbing about 16% of its weight of water at 50% relative humidity and being reactivated by heating at about 250 °C (482 °F).

activated carbon—family of carbonaceous substances manufactured by processes that develop their adsorptive properties. Includes activated charcoal.

activated complex—high-energy intermediate species produced in chemical reactions as a result of the collision of reactant molecules.

activated silica—aqueous alkali silicate solution that has been partially neutralized to become colloidal and somewhat polymerized. Used as a flocculant assisting the alum treatment of waters for color removal and for improving the softening of organic-containing waters.

activated sludge process—aerobic biological process whereby soluble organic matter is converted to solid biomass for removal by gravity or filtration.

activation—changing of a passive metal surface to a chemically active state. *Contrast with* **passivation**.

activation analysis—analytical technique capable of detecting most elements when present in a sample of milligram quantities or less. Activation analysis employs neutrons, protons, or alpha particles as the activating flux.

activation control—term used to describe the control of a corrosion process, i.e., whether it is either the anodic or the cathodic charge-transfer reaction that controls the overall corrosion rate.

activation energy—minimum energy needed to produce an activated complex. Also the minimum energy required for a reaction to occur.

activation pickling—acid treating of metals to remove light surface oxide prior to enameling, tinning, galvanizing, or electroplating to assist adhesion of the surface coating.

activation polarization electrochemical polarization in which the electrochemical reaction is controlled by a step in the reaction sequence requiring a specific energy be overcome for it to proceed at the metal-electrolyte interface. *See also* **polarization**.

activator—material that causes an accelerator or catalyst to begin its function.

active—negative direction of electrode potential. Also a state in which a metal tends to corrode. *Contrast with* **passive**.

active cutting oils—cutting oils containing chemically active ingredients, usually sulfur, chlorine, or phosphorus additives, to improve their boundary lubrication.

active mass—number of gram-molecular weights of a substance per liter of solution.

active metal—metal with a high tendency to react or corrode.

active potential—potential of a corroding material.

active solvent—solvent whose primary function is to dissolve the resin constituent of a paint. Examples include ketones, esters, and glycol ethers. *See also* **latent solvents**.

activity—thermodynamic function used in place of concentration in equilibrium constants for reactions involving nonideal gases and solutions. It is the ion concentration corrected for deviations from ideal behavior by multiplying it by the respective activity coefficient.

activity coefficient—characteristic of a quantity expressing the deviation of a solution from ideal thermodynamic behavior and often used in connection with electrolytes.

acute toxicity—any direct lethal action of one or more substances to a living test organism that is demonstrable within 96 hours or less.

acyclic—organic compound that does not have a ring in its molecule.

acyl groups—radicals derived from carboxylic acids by removal of the hydroxyl group, i.e., *RCO*–.

addition agent—electroplating term for material added in small quantities to a plating solution for the purpose of modifying the character of the deposit.

addition polymerization—polymerization in which monomers are linked together without the splitting off of water or other simple molecules and involves the opening of a double bond.

addition reaction—reaction in which two or more reactants are chemically combined to produce one product.

additive—any substance added in small quantities to another substance or material to modify and, usually, improve its properties.

adduct—compound formed by an addition reaction.

adherend—body held to another body by an adhesive.

adhesion—condition in which two surfaces are held together by interfacial forces. These may consist of valence forces, van der Waals forces, interlocking physical action, or some combination of all three. *Compare with* **cohesion**.

adhesion, coating—adhesion of a coating to a substrate is classified as belonging to one or more of three processes: mechanical, polar, or chemical adhesion.

adhesive—substance capable of holding materials together by surface attachment.

adhesive failure—separation of two bonded surfaces that occurs within the bonding material.

adhesive wear—wear due to localized bonding between contacting solid surfaces leading to material transfer between the two surfaces or material loss from either surface.

admiralty metal—copper-base alloy whose nominal composition is 71% copper, 28% zinc, and 1% tin with smaller amounts of antimony and arsenic also usually present. *See also* **yellow brass**.

adsorbate—any substance that is or can be adsorbed.

adsorbed water—water held on the surface of a material by electrochemical forces. Its physical properties are substantially different from those of absorbed water or chemically combined water.

adsorbent—solid material on the surface of which adsorption occurs.

adsorption—formation of a layer of gas, liquid, or solid on the surface of a solid. There are two types, depending on the nature of the forces involved. In chemisorption, a single layer of molecules, atoms, or ions is attached to the surface by chemical bonds and is essentially irreversible. In physical adsorption, attachment is by the weaker van der

Waals forces whose energy levels approximate those of condensation. *Compare with* **absorption**.

adsorption inhibitor—corrosion inhibitor that is usually an organic compound that adsorbs on a metal surface to suppress metal dissolution and reduction reactions. In most cases, adsorption inhibitors affect both the anodic and cathodic processes, although the effect may be unequal. Organic amines are the most common adsorption type inhibitors. *See also* **corrosion inhibitor, passivation inhibitor**, and **precipitation inhibitor**.

AE—*abbreviation for* **acoustic emission testing**.

AEC—abbreviation for the U.S. Atomic Energy Commission.

aeration—exposing to the action of air. Also the action of causing air to pass into or through a system.

aeration cell—*see* **oxygen concentration cell** and **differential aeration cell**.

aerobic—ability to take place in the presence of oxygen gas. *Opposite of* **anaerobic**.

aerobic bacteria—organisms requiring molecular oxygen for respiration.

aerosol—colloidal dispersion of solid or liquid particles in gaseous media.

AES—*abbreviation for* **Auger electron spectroscopy**. Also the abbreviation for the American Electroplaters' Society.

AFNOR—abbreviation for the Association Française de Normalisation (French).

AFS—abbreviation for the American Foundrymen's Society.

afterboiler—section of the boiler system that includes the superheater, turbines, steam handling equipment, and condensate lines.

afterboiler corrosion—corrosion in steam-generating equipment as manifested by increased maintenance to replace or repair piping and equipment in the steam and condensate system resulting from oxygen attack and low pH due to formation of carbonic acid from the decomposition of water carbonate and bicarbonate to carbon dioxide.

agar (agar-agar)—polymers of galactose hydrocolloid taken from sea algae plants and that are capable of gelling.

age hardening—hardening of a metal or alloy by aging, usually after rapid cooling or cold working.

age resistance—resistance to deterioration with time.

agglomerate—fine particles gathered together into a larger mass.

aggregate—inert granular material, such as sand, gravel, crushed stone, vermiculite, perlite, or crushed slag, that, when bound together with a cement matrix, forms concrete or mortar.

aggressive index (AI)—measure of water corrosivity defined by AWWA Standard C-400 as follows: $AI = pH + \log[(A)(H)]$ where pH is the hydrogen ion concentration in pH units, A is the total alkalinity in milligrams per liter as $CaCO_3$, and H is the calcium hardness in milligrams per liter as $CaCO_3$. $AI < 10$ is very aggressive (corrosive), $AI = 10$ to 12 is moderately aggressive, and $AI > 12$ is nonaggressive.

aging—process of exposing materials to an environment for an extended and, usually, specified interval of time to induce change in properties. Also a change in properties of certain metals and alloys occurring at ambient or moderately elevated temperatures after their hot working, heat treatment, or cold working.

agricultural land—land used for range crops, pastures, food, feed and fiber crops.

AI—*abbreviation for* **aggressive index**.

AIA—abbreviation for the American Institute of Architects.

AICE—abbreviation for the American Institute of Chemical Engineers.

AIHC—abbreviation for the American Industrial Health Council.

air—earth's atmosphere. Uncontaminated dry air at sea level is composed of 78.08% nitrogen, 20.95% oxygen, 0.93% argon, 0.03% carbon dioxide, 0.0018% neon, 0.0005% helium, 0.0001% krypton, and 0.00001% xenon. In addition to water vapor, air in some localities may contain gaseous sulfur compounds, hydrocarbons, hydrogen peroxide, ozone, and dust particles among other contaminants. *See also* **atmosphere**.

air bumping—injection of shots of air into heat exchangers to disturb normal water flow patterns and cause insulating deposits to break loose.

air conditioning—controlling various physical and chemical conditions of the prevailing atmosphere such as temperature, humidity, motion, distribution, dust, bacteria, odor, and toxic gases.

air contaminant—any undesired or harmful substance of either man-made or natural origin in the ambient air, e.g., particulates, mists, fumes, etc.

air content—volume of air voids in cement paste, mortar, or concrete, exclusive of pore space in aggregation particles, and usually expressed as a percentage of total volume of the material.

air-drying—describes paints or varnishes that dry by oxidation or evaporation on simple exposure to air and without aid of heat or a catalyst.

air emissions—release or discharge of pollutants into the ambient air.

air gas—*see* **producer gas**.

air jet—sandblasting gun in which abrasive is conveyed to the gun by a partial vacuum.

airless spraying—spraying by using hydraulic pressure instead of air for the atomization. Also paint application system in which paint under high pressure is passed through a nozzle and broken down into droplets when it enters the lower pressure region outside the nozzle tip.

air pollutant—dust, fumes, mist, smoke, and other particulate matter, vapor, gas, or odorous substances released or discharged into the ambient air.

air-rumbling—periodic injection of air into a water circuit to maintain its solids in suspension.

air sampling—determining types and quantities of atmospheric contaminants by measuring and evaluating a representative sample of air.

air stripping—remediation technique in which an upward-moving air stream is passed through water to blow contaminants out.

air washers—equipment used to clean air and either heat it in the winter or cool it in the summer. They are spray chambers through which air is passed. Greatest use is in the textile industry during textile processing.

AISI—abbreviation for the American Iron and Steel Institute.

AISI classification system—abbreviation for the American Iron and Steel Institute's classification system of the composition of carbon and alloy steel by using a four or five digit number that is generally based on the percentage of carbon and other alloying elements in the metal.

alclad—wrought composite aluminum product comprised of an aluminum alloy core surfaced on one or both sides with a metallurgically bonded aluminum alloy coating that is anodic to the core and thus protects the core from corrosion.

alcohol engine coolant concentrate—engine coolant concentrate whose base is essentially methanol (wood alcohol) or ethanol and that also incorporates soluble corrosion inhibitors. Alkalinity agents, antifoamants, and characterizing coloring dye may also be incorporated.

alcohols—organic compounds characterized by the presence of one or more hydroxyl groups (–OH). Used in the manufacture of synthetic resins, rubbers, and plasticizers.

aldehyde resin—synthetic resin made by reacting various aldehydes with condensation agents such as phenol, urea, aniline, and melamine.

aldehydes—class of organic compounds having a generic formula RCHO and characterized by a carbonyl group (RC=O).

algae—simple plants containing chlorophyll that are often microscopic in size and found in both sea and fresh waters. Under favorable conditions, they may grow in colonies to produce large masses.

algicide—chemical agent used to destroy algae.

alicyclic—organic compound containing a ring of atoms and is aliphatic, e.g., cyclohexane (C_6H_{12}).

aliphatic hydrocarbons—straight chain organic compounds that are either alkanes, alkenes, alkynes, or their derivatives. All noncyclic organic compounds are aliphatic and cyclic aliphatic compounds are alicyclic.

aliphatic solvents—hydrocarbon solvents composed primarily of aliphatic compounds and which may include paraffinic and cycloparaffinic (naphthenic) compounds.

alkali—base that dissolves in water to yield hydroxide (OH^-) ions.

alkali metals—metals of group IA of the periodic system, i.e., lithium, sodium, potassium, rubidium, cesium, and francium.

alkalimetry—volumetric analysis to determine amount of acid present by titrating with a standard solution of alkali.

alkaline—describing a solution that has an excess of hydroxide (OH^-) ions, i.e., having a pH greater than 7.

alkaline cleaning—immersion or spray process for removing soil, e.g., oil, grease, waxy solids, metallic particles, dust, carbon particles, or silica, from metallic surfaces by their emulsification, dispersion, saponification, or any combination of these using an alkaline cleaner. Alkaline cleaners typically are comprised of a builder (alkali metal orthophosphate, condensed phosphate, hydroxide, silicate, carbonate, bicarbonate, or borate), one or more organic or inorganic additives, e.g., glycols, glycol ethers, chelating agents, or polyvalent metal salts, to promote cleaning, and one or more surfactants (anionic, cationic, nonionic, or amphoteric) for detergency, emulsification, and wetting.

alkaline derusting—dissolution of rust from ferrous surfaces using hot, strongly alkaline aqueous formulations containing chelating agents such as EDTA and gluconates. Dissolution can be accelerated by application of an anodic or periodic reversal electric current.

alkaline earth metals—metals of group IIA of the periodic table, i.e., beryllium, magnesium, calcium, strontium, barium, and radium.

alkalinity—quantitative capacity of an aqueous media to react with hydrogen ions. The alkalinity of industrial waters is due mainly to the presence of bicarbonate, carbonate, and hydroxyl ions.

alkali resistance—degree to which a material resists reaction with alkaline substances such as aqueous alkaline solutions, lime, cement, plaster, soap, etc.

alkali swelling—gel formation in concrete with resultant swelling that causes a reduction in its strength or, possibly, its complete destruction. Initiated by the water absorption of poorly crystallized ballast or amorphous silica.

alkane—hydrocarbon that contains only carbon-carbon single bonds; a saturated hydrocarbon.

alkanes—saturated hydrocarbons of the general formula C_nH_{2n+2}. *See also* **paraffin**.

alkene—hydrocarbon that contains at least one carbon-carbon double bond; an unsaturated hydrocarbon.

alkenes—unsaturated hydrocarbons containing one or more double carbon-carbon bonds. *See also* **olefins**.

alkyd resin—generic name for family of synthetic resins prepared by reacting polyhydric alcohols with polybasic acids to form a polyester convertible into a cross-linked form. Modified alkyds substitute, in part, a monobasic acid such as vegetable fatty acid for the polybasic acid.

alkyl benzene sulfonate (ABS)—detergent produced by sulfonating a detergent alkylate, i.e., a ring substituent of an alkyl group sufficiently large to confer detergent properties.

alkyl groups—monovalent aliphatic radicals derived from aliphatic hydrocarbons by removal of a hydrogen atom, e.g., CH_3^- (the methyl group).

alkynes—unsaturated hydrocarbons containing one or more triple carbon-carbon bonds.

alligatoring—surface cracking of an organic coating film to have an appearance similar to alligator hide. A special form of checking in which the surface hardens and shrinks at a much faster rate than the body of the coating. *See also* **checking**.

allotonic—additive that changes the surface tension of water or an aqueous solution.

allotropy—existence of an element in two or more different forms having different characteristics, e.g., oxygen and ozone; carbon black, diamond, and graphite.

alloy—microscopically homogeneous metallic mixture or solid solution of two or more elements to produce one or more phases. Alloys are either homogeneous, where the components are completely soluble in one another and being of only one phase, or they are heterogeneous, where the alloy is a mixture of two or more separate phases. The term alloy is also used to describe resin, polymer, and plastic mixtures formed from two or more immiscible polymers united by another component and having enhanced performance properties.

alloy plating—codeposition of two or more metallic elements.

alpha iron—body-centered cubic form of pure iron stable below 910 °C (1670 °F).

alpha particle (alpha ray)—one of the particles emitted in radioactive decay. It is a high energy helium nucleus (He^{2+}) consisting of two protons plus two neutrons bound together.

alternate immersion test—exposure of a material in frequent, perhaps fairly long, immersion in either fresh or salt water alternated with exposure to the atmosphere above the water.

alternating current (a.c.) (AC)—electric current that reverses its direction with a constant frequency. *Compare with* **direct current**.

alum—hydrated aluminum sulfate ($Al_2(SO_4)_3 \cdot 18H_2O$). *See also* **alums**.

alumina—very-high temperature refractory material (super refractory) composed of sintered aluminum oxide (Al_2O_3) used as a ceramic lining material exhibiting good corrosion and wear resistance and a high hardness.

aluminizing—forming of aluminum or aluminum alloy coating on a metal by a hot dipping, hot spraying, or diffusion process.

aluminum brass—copper aluminum alloy containing up to 5% aluminum.

aluminum bronze—copper aluminum alloy containing from 5 to 11% aluminum.

aluminum paint—coating consisting of a mixture of metallic aluminum pigment in powder or paste form dispersed in a suitable vehicle.

aluminum paste—metallic aluminum flake pigment in paste form, consisting of aluminum, solvent, and various additives.

aluminum transport deposition—problematic term applicable to vehicular liquid cooled engine systems utilizing heat-rejecting aluminum surfaces (cylinder head or block). Corrosion of these surfaces followed by deposition of insoluble aluminum salts in the radiator section of the system can result in loss of heat-exchange capacity and possible engine overheating.

alums—group of double salts with the formula $A_2SO_4 \cdot B_2(SO_4)_3 \cdot 24H_2O$, where A is a monovalent metal and B is a trivalent metal. The originally described alum contains potassium and aluminum sulfate (called potash alum).

Alzak process—proprietary electrolytic brightening process for polished aluminum. Accomplished by anodic smoothing in an aqueous fluoboric acid electrolyte.

amalgam—solution alloy of mercury with one or more other metals.

ambient—surrounding environmental conditions, such as temperature or pressure.

ambient air quality—average atmospheric purity, as distinguished from discharge measurements taken at the source of some air pollution. A description of the general amount of pollution present in a broad area.

ambient temperature—commonly, but incorrectly, used to describe room temperature. Correctly, it is the temperature of the surrounding medium coming into contact with a material or apparatus. *See also* **room temperature**.

amendment—any process changing or correcting a system for the better with regard to its corrosion resistance or protection.

amides—organic compounds containing the group $-CO \cdot NH_2$. Used as a curing agent when combined with epoxy resin.

amine cracking—stress-corrosion cracking of carbon steel by aqueous amine solutions, which solutions are often used to remove hydrogen sulfide and carbon dioxide from refinery and petrochemical plant streams.

amines—organic compounds derived by replacing one or more of the hydrogen atoms in ammonia by organic groups. Primary amines have one hydrogen replaced; secondary amines have two hydrogens replaced; and tertiary amines have all three hydrogens replaced (RNH_2, R_2NH, R_3N, respectively). Used as a curing agent when combined with epoxy resin.

amine salt—salt that results when an amine combines with an acid. The general formula for a primary amine salt is $RNH_3 + X^-$.

amino acid—any of a group of water-soluble organic compounds that possess both a carboxyl ($-COOH$) and an amino ($-NH_2$) group attached to the alpha-carbon atom.

amino resin—resin made by polycondensation of a compound containing amino groups, such as urea or melamine, with an aldehyde, such as formaldehyde, or an aldehyde-yielding material.

ammeter—instrument for measuring the magnitude of electric current flow.

ammine—coordination complex in which the ligands are ammonia molecules.

ammonia—colorless compound NH_3, which is gaseous at room temperature, exhibits a characteristic strong pungent odor, and is a weak base. It readily dissolves in water to yield a weakly basic solution (aqueous ammonia).

ammonia attack—*see* **condensate corrosion inhibitors**.

ammoniacal—solution in which the solvent is aqueous ammonia.

ammonium hydroxide (aqueous ammonia)—term commonly given to an aqueous solution of ammonia that contains some ammonium and hydroxide ions (NH_4^+ and OH^-, respectively). Because it mostly retains its nitrogen atoms in the form of NH_3, aqueous ammonia is a better name.

ammonium ion—monovalent cation NH_4^+.

amorphous—being noncrystalline or devoid of regular structure.

amorphous alloy—semiprocessed alloy produced by a rapid quenching and direct casting process resulting in metal with noncrystalline structure. *Sometimes called* **metallic glass**.

AMP—abbreviation for 2-amino-2-methyl-1-propanol; used as a dispersant, corrosion inhibitor, or pH modifier.

ampere—symbol A. SI unit of electric current. It is defined as the constant current that, maintained in two straight parallel infinite conductors of negligible cross section placed one meter apart in a vacuum, would produce a force between the conductors of 2×10^{-7} Nm^{-1}.

ampholyte ion—*see* **zwitterion**.

amphoteric—substance capable of reacting either as a weak acid or weak base.

amphoteric oxides—metal oxides, e.g., aluminum, zinc, and iron oxides, capable of reacting with alkalies to give soluble complex cations; Al_2O_3 forms $2AlO_2^-$ (aluminate anion) above pH 8.6, ZnO forms $HZnO_2^-$ (bizincate anion) above pH 10.5, FeO forms $HFeO_2^-$ (bihypoferrite anion) above pH 12.2.

amplifier—device for increasing the strength of an electrical signal by drawing energy from a separate source to that of the signal.

amplitude—maximum value of a periodically varying quantity.

AMS—abbreviation for Aerospace Material Specification (of SAE).

anaerobic—free of air or uncombined oxygen. Also the ability to take place without the presence of oxygen gas. *Opposite of* **aerobic**.

anaerobic corrosion—corrosion of iron or steel under anaerobic moist conditions, usually caused by the sulfide metabolic reaction products (biogenic sulfides) of sulfate-reducing bacteria (SRB). It usually occurs where there is an abundance of sulfate and the metal substrate environment is between pH 5.5 and 8.5. The sulfide produced retards cathodic reactions and accelerates anodic dissolution to precipitate ferrous sulfide (FeS). *See also* **sulfate-reducing bacteria**.

anaerobic organism—organism that can thrive in the absence of oxygen.

analysis—determination of the chemical components in a sample. Qualitative analysis refers to the identification of one or more of the components of a sample. Quantitative analysis refers to the determination of the quantity of one or more of the components of a sample.

anchorite—zinc-iron phosphate coating for iron and steel.

anemometer—instrument for measuring the speed of wind or other flowing fluid.

angle blasting—blast cleaning at angles less than 90°. *See also* **blast angle**.

angstrom—unit of length equal to 10^{-10} meter.

anhydride—compound derived from an inorganic or organic acid by the elimination of water.

anhydrite—mineralogical name for anhydrous calcium sulfate.

anhydrous—dry and free of water in any form.

animal fats and oils—organic products derived from the fatty tissue of such animals as cattle, sheep, and swine. The oils are of unsaturated chemical structure and are liquid at room temperature. The fats are saturated and are solid or semisolid at room temperature. Both are often used as boundary lubricants in cutting and grinding operations.

anion—negatively charged ion.

anion exchange material—material capable of the reversible exchange of negatively charged ions.

anionic emulsion—emulsion in which the emulsifying system produces a predominance of negative charges on the discontinuous phase.

anionic surfactant—surfactant giving negatively charged ions in aqueous solution.

anisotropic—material having different values for a property relative to its different directions. Opposite of isotropic.

annealing—process of heating a metal or metal alloy to and holding at a suitable temperature followed by cooling at a suitable rate. Used primarily to soften metallic materials, relieve them of stress, and to produce desired changes in their microstructure. The term annealing is also applied to the heat treatment of polymer alloys to effect similar benefits.

anode—electrode of an electrolytic cell where oxidation is the principal reaction. Also the electrode where corrosion usually occurs and from whence metal ions enter solution.

anode corrosion—dissolution of a metal acting as an anode.

anode corrosion efficiency—ratio of the actual corrosion of an anode to the theoretical corrosion (by weight loss) as calculated by Faraday's law from the quantity of electricity that has passed.

anode effect—effect produced by polarization of the anode in electrolysis characterized by a sudden increase in voltage and decrease in amperage due to the anode being separated from the electrolyte by a produced gas film.

anode efficiency—current efficiency at the anode.

anode energy content—term used in describing sacrificial anodes. It is the characteristic quantity of electricity (in ampere hours) that can be provided for anodic protection purposes by one pound of a given sacrificial anode metal. It depends on the electrochemical equivalent of the metal used and its efficiency in operation.

anode film—layer of solution in contact with the anode that differs in composition from that of the bulk solution; or the outer layer of the anode consisting of oxidation or reaction products of the anode metal.

anode, sacrificial—chemically active metal which, when electrically connected, will provide the energy needed to cathodically protect a less anodic metal. Zinc, aluminum, and magnesium are commonly used as sacrificial anodes.

anodic cleaning—electrolytic cleaning in which the workpiece is made the anode. Also called reverse-current cleaning. *Contrast with* **cathodic cleaning**.

anodic coating—protective, decorative, or other functional coating formed on a metallic surface by its chemical conversion in an electrolytic oxidation (anodic) process. Also the coating formed on aluminum and aluminum alloys by an exothermic reaction on their surface with nascent atomic oxygen produced by the electrolysis of an aqueous electrolyte, usually sulfuric, chromic, phosphoric, or oxalic acid, so as to form a continuous film of hexagonal aluminum oxide columns having a central capillary capable of absorbing dyestuff, sealants, lubricating oil, etc.

anodic electrodeposition—electrodeposition in which the substrate is the anode (positive in charge). In painting, the paint is made cathodic or negative in charge.

anodic inhibitor—chemical substance or combination of substances that prevent or reduce the rate of the anodic or oxidation reaction (corrosion) by physical, physicochemical, or chemical action.

anodic passivation—metal passivation occurring when polarized anodically in an appropriate electrolyte.

anodic polarization—change of electrode potential in the noble (positive) direction due to current flow.

anodic protection—technique to reduce the corrosion rate of a metal by polarizing it into its passive region.

anodic reaction—electrode reaction equivalent to a transfer of positive charge from the electronic to the ionic conductor. Anodic reaction is an oxidation process as in the corrosion of metal.

anodized aluminum—aluminum having increased anodic corrosion resistance as a result of being subject to anodizing.

anodized coating—heavy oxide film produced on aluminum by anodizing. *See also* **anodizing**. Titanium, zirconium, and stainless steel are also subject to anodizing and formation of anodized coatings.

anodizing—commercial term for the electrolytic treatment of metals that form stable films or coatings on the surface of the metal. The coating has desirable protective, decorative, or functional properties. Also method of coating aluminum with a protective oxide film by making it the anode in an electrolytic bath containing an oxidizing electrolyte. Titanium, zirconium, and stainless steel are also subject to anodizing. *See also* **chemical coating** and **electrochemical conversion coating**.

anolyte—electrolyte adjacent to the anode in an electrolytic cell. *Contrast with* **catholyte**.

ANSI—abbreviation for the American National Standards Institute, Inc. It is the U.S. member of the ISO.

anticarbonation coatings—coatings for concrete that provide a barrier to carbon dioxide penetration but allow passage of moisture. Used to prevent deterioration of concrete and reinforced concrete due to carbonation. *See also* **carbonation**.

anticondensation paint—coating formulated to minimize effects of condensation of moisture under intermittently dry and humid conditions. The coatings are normally possessive of a matte-textured finish and often contain a heat-insulating filler.

anticorrosion paint—paint used to prevent or slow the corrosion of metals. Most often used on iron or steel.

antidegradant—compounding material incorporated in plastics or in paint coatings to retard their deterioration due to oxidation, light, ozone, or any combination of these.

antifoaming agent—additive to liquids to eliminate or reduce their tendency to foam.

antifouling—preventing or resisting marine organism attachment or growth on a submerged surface. Usually achieved by application of a toxic coating layer on the surface of the submerged material.

antifreeze agent—substance that is added to a liquid, usually water, to lower its freezing point. Ethylene glycol is the most commonly employed antifreeze used in automotive and truck engine cooling systems. In addition to the ethylene glycol, vehicular antifreezes usually contain small concentrations of soluble additives to retard foaming, control pH, and inhibit corrosion. The term antifreeze is also applicable to the resulting aqueous solution of the diluted ethylene glycol when used as a coolant in the vehicle cooling system. Antifreeze agents are also used in fuels where severe environmental conditions are encountered.

antinucleation—process by which the formation of a nucleus is prevented or inhibited.

antioxidant—oxidation inhibitor. Additive, usually organic, that prevents or reduces the propensity of materials to chemically oxidize or weather. Used in various types of materials, such as rubber, natural fats and oils, food products, gasoline and lubricating oils, to retard their oxidative deterioration, rancidity, gum formation, reduction in shelf life, etc.

antiozonant—compounding additive for elastomers and resins to retard their deterioration caused by ozone.

antipitting agent—addition agent for the specific purpose of preventing gas pits in an electrodeposit. *Also known as* **wetting agent**.

antirust—chemical and emulsifiable oil-type substances added to water when used as an engine coolant to prevent or retard corrosion of the cooling system.

antiscalant—substance added to boiler waters to reduce scaling or alter the nature of scale. Antiscalants are generally classified as either those that chemically react stoichiometrically with water impurities to change their chemical structure (carbonates, phosphates, and chelants) or those that alter the action of the impurities (polymers and threshold acting sequestrants).

antiseptic—agent used to destroy or inhibit the growth of microorganisms.

ant nest corrosion—localized corrosion seen in copper tubes used in air-conditioning units and refrigerators and appearing as an underlying labyrinth of interconnecting channels containing porous copper oxide in random micropaths whose morphology is similar to that of an ant nest when viewed in cross section. Thought to be induced by carboxylic acids produced by hydrolysis of chlorinated organic solvents and lubricating oils.

AOCS—abbreviation for the American Oil Chemists Society.

API—abbreviation for the American Petroleum Institute. Sets specifications for virtually all categories of equipment used in the oil industry.

API gravity—index of specific gravity for crude oils as defined by the American Petroleum Institute. It is inversely related to specific gravity by the equation API gravity = 141.5/specific gravity − 131.5.

apparent density—mass per unit volume of a loose material, such as a powder. The volume includes space occupied by both the particles and the air voids between particles. *Also known as* **bulk density**.

appliance coatings—thermoset coatings used for appliance finishing that are characterized by their hardness, mar resistance, and good chemical resistance.

approximate value—value nearly but not exactly correct or accurate.

aqua regia—mixture of concentrated nitric acid and concentrated hydrochloric acid in the approximate volume ratio of 1:3, respectively. Also known as "royal water."

aquatic—in or on water.

aqueous—pertaining to water. Water-containing or water-based. A solution in which water is the solvent or a cosolvent.

aqueous cleaning—process for metal surface cleaning and its preparation for subsequent surface operations using an aqueous cleaning composition usually containing a surfactant, alkaline salts, and sequestering agent.

aqueous organisms—multitude of animal and plant life, including barnacles, mussels, algae, etc., found in both fresh and salt water, which can attach themselves to solid surfaces during their growth cycle and whose accumulation can cause crevice corrosion and fouling.

aquifer—subsurface geological structure carrying or holding water.

ARBA—abbreviation for the American Railway and Bridge Association.

architectural coatings—protective and decorative coatings intended for on-site application to the interior or exterior surfaces of residential, commercial, institutional, or industrial buildings. *Contrast with* factory-applied **industrial coatings**.

argillaceous—clay-containing.

Armco iron—trademark for a commercially pure iron that is relatively weak and not used where strength is a major requirement.

aromatic compound—organic compound that contains a benzene ring in its molecule.

aromatic solvents—hydrocarbon solvents composed primarily of aromatic hydrocarbon compounds, e.g., benzene, xylene, toluene.

arrest marks—*see* **beach marks**.

Arrhenius theory of electrolytic dissociation—states that the molecule of an electrolyte can give rise to two or more electrically charged atoms or ions.

artesian aquifer—*see* **confined aquifer**.

artesian well—well drilled through impermeable strata to reach water capable of rising to the surface by internal hydrostatic pressure.

artificial aging—aging above room temperature.

artificial weathering—exposure to cyclic laboratory conditions involving changes in temperature, humidity, and radiation in an attempt to produce changes in a material in a short time similar to those observed after longer-term exposure to outdoor or natural weathering.

aryl—pertaining to monovalent aromatic groups, such as phenyl (C_6H_5-).

asbestos—group of natural fibrous impure silicate materials. Applied to six naturally occurring minerals: chrysotile, amosite, crocidolite (blue asbestos), anthophyllite, tremolite, and actinolite.

ASCC—abbreviation for the accelerated simulated can corrosion test. See ASTM STP 866 for details.

ash—any inorganic residue remaining after ignition of a combustible substance.

ASLE—former abbreviation for the American Society of Lubrication Engineers. Now known as the Society of Tribologists and Lubrication Engineers (STLE).

ASM—abbreviation for the former American Society for Metals. Now known as ASM International.

ASME—abbreviation for the American Society of Mechanical Engineers.

asperity—small scale protuberance of a solid surface.

asphalt—black to dark brown residue from petroleum refining. Also a natural complex hydrocarbon whose predominating constituents are bitumens.

asphalt cut back—asphalt with thinner. Also an asphalt solution and an asphalt coating formed from dissolved asphalt.

asphaltene—high-molecular-weight hydrocarbon fraction precipitated from asphalt by a designated paraffinic naphtha solvent at a specified temperature and solvent-asphalt ratio.

asphaltic—being essentially composed of or similar to asphalt. Frequently applied to lubricating oils derived from crudes that contain asphalt.

asphalt mastic—dense mixture of sand, crushed limestone, and fiber bound with air-blown asphalt. Used as a filler, stopper, putty, adhesive, or high-build coating.

asphalt paint—liquid asphaltic product containing pigments or fillers.

association—combination of molecules of one substance with those of another to form chemical species that are held together by forces weaker than normal chemical bonds, e.g., ethanol and water to form a mixture (associated liquid) in which hydrogen bonding holds the different molecules together.

A-stage—early stage in the preparation of certain thermosetting resins in which the material is still soluble in certain liquids, and may be liquid or capable of becoming liquid upon heating. *Compare with* **B-stage** and **C-stage**.

ASTM—abbreviation for the American Society for Testing and Materials. It issues specifications, methods of testing, recommended practices, definitions, and tentative specifications.

atmosphere—whole mass of air surrounding the earth and being composed largely of oxygen and nitrogen. *See also* **air**. *Also called* **troposphere**. Also used to describe a specific gaseous mass containing any number of constituents and in any proportion produced by man for special purposes. Also a unit of pressure equal to the pressure exerted by a vertical column of mercury 760 millimeters high at a temperature of 0 °C under standard gravity.

atmospheric corrosion—gradual degradation or alteration of a material by contact with substances present in the atmosphere, e.g., oxygen, carbon dioxide, water vapor, and sulfur or chlorine compounds.

atmospheric dust—particulate matter found in the ambient atmosphere. Tends to increase corrosion of metallic surfaces exposed to it, particularly when combined with moisture. Industrial atmospheres typically carry suspended particles of carbon and carbon compounds, metal oxides, sodium chloride, ammonium, and other salts. Marine atmospheres typically contain alkali and alkaline earth metal salt particles.

atmospheric pressure—pressure exerted by the atmosphere at any specific location. It is approximately 14.7 pounds per square inch or 760 millimeters of mercury at sea level.

atom—neutral particle composed of protons, neutrons, and electrons and that is the smallest part of an element that can enter into chemical combination. *See also* **Bohr model (of the atom)**.

atomic absorption spectroscopy (AAS)—analytical technique in which a sample is vaporized and the nonexcited atoms absorb electromagnetic radiation at characteristic wavelengths.

atomic emission spectroscopy (AES)—analytical technique in which a sample is vaporized and the atoms present are detected by their emission of electromagnetic radiation at characteristic wavelengths.

atomic mass—average mass of the naturally occurring isotopes of an element compared to ^{12}C. Traditionally called atomic weight. *See also* **atomic weight**.

atomic number (proton number)—number of protons in the nucleus of an atom. Symbol Z. Atomic number is equal to the number of electrons orbiting the nucleus in a neutral atom and, therefore, the positive charge on the nucleus of an atom.

atomic weight—relative atomic mass of an atom compared to other atoms. The relative weights are usually expressed in atomic mass units (amu) which unit is equal to $1/12$ of the mass of an atom of isotope ^{12}C.

attapulgite clay—$(Mg,Al)_5Si_8O_{20}(OH)_2 \cdot 8H_2O$. A colloidal, viscosity-building clay principally used in the salt-water muds of oil and gas well drilling.

attrition—process to reduce the size of various substances usually by a rubbing action that is accompanied by shear and impact forces.

Auger effect—ejection of an electron from an atom without the emission of an X- or gamma-ray photon as a result of the de-excitation of an excited electron within the atom. Principal use is in the Auger analysis of metal surfaces.

Auger electron spectroscopy—chemical analysis of a surface zone composed of elements of atomic number above 2 and determination of its thickness by irradiating the surface with electrons and determining the emission of Auger electrons. It is used to study sample surfaces to a depth of about 0.5 to 2 nanometers.

austempering—rapid cooling of steel to below the "nose" of the temperature-time-transformation curve to transform austenite to bainite, which is a ductile structure.

austenite—face-centered-cubic solid solution of carbon or other elements in gamma iron.

austenitic stainless steels—stainless steels having two principal alloying elements, chromium and nickel, and often also manganese, and possessing a nonmagnetic microstructure consisting essentially of austenite. They have the highest general corrosion resistance of any of the stainless steels. Austenitic stainless steels constitute the AISI type "200" and "300" series, with the most common being type 304.

autocatalytic plating—deposition of a metal coating by controlled chemical reduction catalyzed by the metal or alloy being deposited. Also known as electroless plating.

autoclave—closed vessel for conducting a chemical reaction or other operation under pressure and heat.

autohesion—bond formed between two contiguous surfaces of the same material that prevents their separation at the place of contact. Also known as self-adhesion.

autophoretic paints—water-reducible paints deposited on metal surfaces by the catalytic action of the metal on the paint materials.

autoxidation—spontaneous oxidation process taking place between matter and molecular oxygen or air at moderate temperatures usually below 150 °C (302 °F) without visible combustion.

auxiliary anode—supplementary anode used in electroplating and positioned so as to raise the current density on a certain area of the cathode to obtain better distribution of plating.

auxiliary electrode—electrode in an electrochemical cell that is used to transfer current to or from a test electrode. Also called counter electrode.

auxiliary solvent—liquid used in addition to the primary solvent and generally used to replace part of the primary solvent for a special effect or as a matter of economics.

average—in a series of observations, it is the total divided by the number of observations. Synonyms include arithmetic average, arithmetic mean, and mean.

Avogadro's law—equal volumes of all gases contain equal numbers of molecules when at the same pressure and temperature. This is true only for ideal gases, but it approximates the condition for most real gases.

Avogadro's constant—the number of atoms or molecules in one mole or a gram-molecular weight of any substance. It is equal to 6.02×10^{23}. Formerly called Avogadro's number.

Avogadro's principle—numbers of molecules present in equal volumes of gases at the same temperature and pressure are equal.

AWS—abbreviation for the American Welding Society.

AWT—abbreviation for the American Water Technologies, Inc.

AWWA—abbreviation for the American Water Works Association.

azeotropic system—mixture of two liquids that boils at constant composition. Also called azeotropic mixture and constant boiling mixture.

B

Babbitt metal—any of a group of related alloys used for making bearings. They consist of tin with about 10% antimony, 1 to 2% copper, and, often, some lead.

babbitting—process by which softer metals, basically Babbitt metal, are bonded chemically or mechanically to a shell or stiffener that supports the weight and torsion of a rotating, oscillating, or sliding shaft to prevent its galling or scoring.

Babo's law—vapor pressure over a liquid solvent is lowered approximately in proportion to the quantity of a nonvolatile solute dissolved in the liquid.

back emf—potential setup in an electrolytic cell that opposes the flow of current and is caused by such factors as concentration, polarization, and electrode films.

backfill—material used to fill the space around anodes, vent pipe, and buried components of a cathodic protection system.

backflushing—*same as* **backwashing**.

backwashing—turbulently reversing the flow of water through a heat exchanger to loosen and remove light silt and mud.

bacteria—microscopic, one-celled organisms that reproduce by fission or by spores and that are identified by their shape: bacillus (rod-shaped), spirillum (curved), and coccus (spherical).

bacterial cleaning—removal of rust and scale by spraying or dipping steel into a solution of an applicable bacterium, inorganic salts, and glucose.

bacterial corrosion—corrosion resulting from substances (e.g., sulfuric acid, ammonia) produced by the activity of certain bacteria. *See also* **bacteriological corrosion, biological corrosion**, and **microbiologically influenced corrosion**.

bactericide—substance used in relatively small quantity to kill bacteria.

bacteriological corrosion—direct or indirect corrosion of buried pipeline brought on by existing microorganisms. Anaerobic bacteria, which are sulfate-reducing, consume hydrogen and cause loss of polarization at the steel pipe surface making attainment of cathodic protection difficult. Aerobic bacteria consume oxygen and oxidize sulfides into sulfates, such as sulfuric acid leading to acid attack. *See also* **bacterial corrosion, biological corrosion**, and **microbiologically influenced corrosion**.

bacteriostat—substance used in relatively small quantity to prevent or slow growth of bacteria.

bag house—mechanical dust collector which is a tubular bag device that can be cleaned on-line while the dust collection process is being performed.

bainite—acicular metastable aggregate of ferrite and cementite resulting from the transformation of austenite at temperatures below the pearlite range but above the martensite start temperature.

Bakelite—proprietary name for certain phenol-formaldehyde resins.

baking—process of drying or curing a coating by the application of heat in excess of 66 °C (150 °F).

baking finish—paint or varnish requiring baking at temperatures above 66 °C (150 °F) for the development of desired properties.

baking soda—sodium hydrogen carbonate ($NaHCO_3$). *Also called* **bicarbonate of soda** and sodium bicarbonate.

ballast tank—tank used to maintain the stability of a ship by adjusting the amount of seawater in the tank.

ball clay—secondary clay characterized by a presence of organic matter, high plasticity, high dry strength, long vitrification range, and light color when fired.

ball mill—rotating cylinder that operates on a horizontal axis in which minerals or ceramic materials are wet- or dry-ground using free-moving pebbles, porcelain, or metallic balls as the grinding media.

bar—c.g.s. unit of pressure equal to 10^9 dynes per square centimeter or 10^5 pascals and that is approximately 750 mm Hg or 0.987 atmosphere.

bare glass—glass, in any of its forms, from which any sizing or finish has been removed, or the term for glass before application of sizing or finish.

barite—*see* **baryte.**

barium metaborate—anticorrosion paint pigment whose nominal composition is $BaB_2O_4 \cdot H_2O$.

bark—layer of a tree outside the cambium comprising the inner and outer bark.

barnacle—any of various marine crustaceans of the order *Cirripedia* that, in the adult stage, form a hard shell and remain attached to a submerged surface.

barometer—device for measuring atmospheric pressure. The common mercury barometer is an open-tube manometer.

barrel (bbl)—standard unit of liquid volume in the petroleum industry equal to 42 U.S. gallons, approximately 35 Imperial gallons, and approximately 160 liters.

barrel coating—painting of large numbers of small pieces such as nails, screws, buttons, etc. by their placement in a barrel with a predetermined weight of coating and rotating to distribute the coating uniformly then following with the injection of drying air while still rotating. Also called tumble coating.

barrel plating (or cleaning)—plating or cleaning in which the work is processed in bulk in a rotating container.

barrier—any material limiting passage through itself of solids, liquids, semisolids, gases, or forms of energy such as ultraviolet light. *See also* **barrier layer**.

barrier coat—coating used to isolate another paint coating from the surface to which it is to be applied to prevent chemical or physical interaction.

barrier layer—usually applied to anodized aluminum to describe the thin, pore-free, semiconducting aluminum oxide region nearest the metal surface and being distinct from the main anodic oxide coating, which has a pore structure. The term also applies to any layer acting as a barrier. *See also* **barrier**.

barrier materials—waterproof and water-vapor-proof wrapping materials for protection of metallics from atmospheric tarnishing and more severe corrosion. Typical barrier materials include films of polyethylene, high- and low-density polyethylene, polyvinyl chloride, linear polyester, waxed or asphaltic-coated paper, and metal foil, or composites of metal foil, plastic, and paper. The wrapping may also enclose a desiccant for added protection.

barrier materials classification—arbitrary classification of barrier materials according to their resistance to passage of liquid water or water-vapor: waterproof meaning a high resistance to passage of liquid water; water-vapor-resistant meaning a water vapor transmission rate (WVTR) no greater than 8 $g^{-2}d^{-1}$ per day at 38 °C (100 °F) from 90 to 2% relative humidity; and water-vapor-proof meaning a WVTR no greater than 1 $g^{-2}d^{-1}$ per day at 38 °C (100 °F) from 90 to 2% relative humidity. *See also* **WVTR**.

barrier pigments—certain pigments, e.g., aluminum, micaceous iron oxide, glass flake, or stainless steel flake, that are used to enhance the corrosion-protective quality of barrier primers.

barrier primer—primer coating that protects the metallic substrate from corrosion by resistance inhibition, i.e., by increasing the electrical resistance between the metal surface and the electrolyte thus eliminating or reducing the corrosion current to a negligible level.

baryte (barite)—natural barium sulfate commonly used for increasing the density of drilling fluids. *Also known as* heavy spar and **blanc fixe**.

base—chemical substance that yields hydroxyl ions (OH⁻) when dissolved in water (Arrhenius definition). A proton acceptor (Lowery-Bronsted definition). An electron pair donator (Lewis definition). *Compare with* **acid**. Also a compound that reacts with a protonic acid to give water and a salt.

basecoat—first coat applied to a substrate. *Also called* **primer, undercoat**, and underneath coat.

base exchange—physicochemical process in which one species of adsorbed ions is replaced by another species.

base metal—metal that readily oxidizes or dissolves to form ions. *Opposite of* **noble metal**.

base units—seven fundamental SI units from which all other SI units are derived. Also units that are defined arbitrarily rather than being defined by simple combinations of other units.

basic—compound that is a base. Also a solution containing an excess of hydroxide, OH⁻, ions. Alkaline.

basic lead chromate—corrosion-inhibitive paint pigment ($PbO \cdot PbCrO_4$).

basic lead silico chromate—corrosion-inhibitive paint pigment produced by forming basic lead chromate in the presence of sand to form a lead chromate coated silica that is subsequently calcined to yield some gamma tribasic lead silicate ($3PbO \cdot PbSiO_2$).

basic salt—compound regarded as being formed by replacing some of the oxide or hydroxide ions in a basic compound by other negative ions, e.g., bismuth(III) chloride oxide (BiOCl).

basic zinc chromate—U.S. term for the corrosion-inhibitive paint pigment $5ZnO \cdot CrO_3 \cdot 4H_2O$. Also known as zinc tetroxy chromate. Also the name used in Europe for zinc yellow. *See also* **zinc yellow**.

basis metal—metal upon which coatings are deposited.

batch treatment—method of adding a quantity of corrosion inhibitor at one time to a system to provide protection for an extended period. Used in treating oil and gas wells and to add protection to automobile cooling systems.

batt—thermal insulation in the form of a blanket rather than loose filling.

battery—electric cells joined together for the purpose of producing an electric current. It is the same as a corrosion cell, but a battery involves a corrosion process deliberately made more efficient, while a corrosion cell is generally an unwanted reaction that one attempts to stop or make less efficient. If the electric cells are rechargeable, the battery is called a secondary or storage battery.

Baumé scale—special hydrometer for determining the specific gravities of liquids. Its scale utilizes Baumé degrees, which can be converted to true specific gravities by reference to conversion tables.

beach marks—macroscopic, irregular elliptical, or semielliptical rings radiating outward from one or more origins marking the progression of fatigue fracture. Also called arrest marks or clamshell marks.

bearing corrosion—chemical attack on bearing metal, or on one of the metals in a bearing alloy, due to acids evolved during chemical deterioration of a lubricating oil or grease.

bedrock—essentially, the continuous body of rock underlying overburdened soils.

Beer's law—if two solutions of the same compound are made in the same solvent, one of which is, e.g., twice the concentration of the other, the absorption due to a given thickness of the first solution should be equal to the absorption at twice the thickness of the second.

beeswax—soft, natural wax with a melting point of 63 °C (145 °F) that is a mixture of crude cerotic acid and myricin, which is obtained by separation from honey.

Beilby layer—altered surface layer formed on a crystalline solid during a wear process or by mechanical polishing.

bentonite—very fine-grained clay that is a montmorillonite mineral mixture (an aluminum silicate clay) as opposed to a compound and that is derived from volcanic ash. Two classes are recognized: sodium bentonite, which swells in water, and calcium bentonite, which exhibits little swelling. Also known as bentonite clay.

benzene—clear, colorless, highly flammable liquid of characteristic odor having a solidification point of 5.5 °C (42 °F) and boiling point of 80 °C (176 °F) whose chemical name is cyclohexatriene (C_6H_6). Obsolete name for benzene is benzol.

BESA—abbreviation for the British Engineering Standards Association.

BET—abbreviation for the Brunauer-Emmett-Teller equation and method for determining the surface area of an absorbent material.

beta particle (beta ray)—one of the particles that can be emitted by a radioactive atomic nucleus. It has a mass about $1/1837$ that of the proton. The negatively charged beta particle is identical with the ordinary electron, while the positively charged beta particle (positron) differs from the electron in having equal but opposite electrical properties. Beams of negatively charged beta particles are used to cure certain paints.

bias—*see under* **statistical terms**.

bicarbonate alkalinity—portion of the total alkalinity in a water that is contributed by bicarbonate ions (HCO_3^-).

bicarbonate of soda—common name for sodium hydrogen carbonate ($NaHCO_3$). *Also known as* **baking soda**.

bifunctional resin—ion-exchange material having both cation and anion exchange sites on the same polymer backbone.

bilge—area in the inside bottom of ships into which liquids are drained.

bimetallic corrosion—corrosion resulting from dissimilar metal contact. *Also known as* **galvanic corrosion**, two-metal corrosion, and dissimilar metal corrosion.

binary acid—acid that contains a hydrogen atom bonded to a nonmetal atom, e.g., HCl.

binary compound—compound that is composed of only two different elements.

binder—resinous adhesive component of a pigmented coating composition. Also the nonvolatile liquid portion of a film.

binder demand—amount of binder needed to completely wet a pigment by displacing all air voids.

bioaccumulation—net accumulation of a substance by an organism as a result of uptake from environmental sources.

bioassay—standardized procedure for the determination of the effects of an environmental variable or substance on living organisms.

biochemistry—science concerned with the study of compounds that are found in living things and the reactions they undergo to sustain life.

biochemical oxygen demand (BOD)—measure of biodegradable organic matter present in water. It is the amount of oxygen taken up by the oxidation of organic matter in water under controlled test conditions, usually involving a 5-day incubation at 20 °C (68 °F).

biocide—chemical used to control the population of troublesome microbes.

biocompatibility—first consideration for materials of any type that are to be used in the body (metallic implants and prosthetic devices). Using corrosion-resistant materials minimizes metal ion release, while incompatible materials can interfere with normal tissue growth near the implant, interfere with systemic reactions of the body, and transfer and deposit metal ions at selective sites or organs.

biodegradable surfactant—surfactant that can be decomposed by biological action.

biodegradation—process by which matter may be broken down by living organisms.

biofouling—undesirable formation of a biological film or growth on an inanimate surface caused by the attachment of macro- and microorganisms with the production of extracellular products.

biological corrosion—not a type of corrosion, but the deterioration of metals as a result of the metabolic activity of microorganisms. *See also* **bacterial corrosion, microbiologically influenced corrosion,** and **sulfate-reducing bacteria**.

biological deposits—deposits of organisms or the products of their life processes.

biological fouling—*same as* **biofouling**.

biomass—any material, excluding fossil fuels, that is or was a living organism and that can be used as a fuel either directly or after a conversion process. Also the total mass of all organisms in a given area.

bioremediation—biological remediation method by which naturally occurring microbes are used to decompose water contaminants.

biosphere—portion of the earth and its atmosphere that can support life.

biota—all animal and plant life of a region or system.

bipolar electrode—electrode in an electrolytic cell that is not mechanically connected to the power supply, but is so placed in the electrolyte between the anode and cathode that the part nearer the anode becomes cathodic and the part nearer the cathode becomes anodic. Also termed intermediate electrode.

bisque—coating of wet process porcelain enamel that has been dried but not fired.

BISRA—abbreviation for the British Iron and Steel Research Association.

bitumen—generic term used to denote any material composed principally of bitumen, i.e., a class of amorphous, black or dark-colored and solid, semisolid, or viscous cementatious substances, natural or manufactured, composed principally of high-molecular-weight hydrocarbons of which asphalts, tars, pitches, and asphaltics are typical.

bituminous coating—coal tar- or asphalt-based coating.

bituminous sands—*see* **tar sands**.

blackening—process for forming a black, blue, or brown oxide basis conversion coating on iron, steel, stainless steel, zinc, or cadmium by immersion in a bath containing, for example, caustic soda or alkaline chromate for purposes of corrosion or abrasion resistance or for color development. *See also* **conversion coating**.

black iron oxide—$FeO \cdot Fe_2O_3$ or Fe_3O_4. Occurs naturally as the mineral magnetite, but can be produced synthetically or as a result of the corrosion of ferrous metal. It is often produced as a finish on iron and steel articles by immersing them in hot oxidizing salts or salt solutions.

black light—light whose energy is just below the visible range of ultraviolet light and mostly of about 365 nanometers. Used on certain penetrating dyes to make them fluoresce in a range visible to the eye. *See also* **penetrant**.

black liquor—liquid material remaining from pulpwood cooking in the soda- or sulfate paper-making process.

black oil—any dark-colored lubricating oil used for the lubrication of machine parts under exposed conditions.

black oxide coating—conversion coating, usually developed on steels, stainless steels, and copper alloys, by immersing the metal in a hot, highly alkaline solution containing suitable oxidizing agents to convert the surface to their oxides.

black phosphate coating—phosphate coating with a matte black finish usually achieved by use of a predip based on antimony or bismuth salts, which form a black smut that becomes incorporated into the phosphate layer.

blanc fixe—barium sulfate ($BaSO_4$).

blank coating (precoating)—process in which metal is prepainted prior to the part being formed.

blast angle—angle of nozzle with reference to the surface. *See also* **angle blasting**.

blast cleaning—cleaning with propelled abrasives.

blast cleaning to visually clean steel—*see* **white blasting**.

blast primer—quick-drying coating applied to blast-cleaned steel to protect the steel from rusting until further coats of paint can be applied.

bleaching—removing color by exposure to sunlight or chemical agents.

bleeding—tendency of a liquid component to separate from a liquid-solid or liquid-semi-solid mixture.

bleedoff—*see* **blowdown**.

blemish—anything marring the appearance of a surface that is not classifiable as a specific defect.

blister cracking—hydrogen-induced pipeline failure originating in steels containing internal flaws by nonmetallic inclusions due to superficial corrosion of the steel by an acid H_2S environment liberating atomic hydrogen, which diffuses into the metal and is released at the inclusion-metal interface as molecular hydrogen under high pressure. *See also* **blistering**.

blistering—bubbles formed in dry or partially dry paint films; most often due to the presence of water or other liquids or gases under the film. Also used to describe a bubble or gaseous inclusion at the surface of a body which, if broken, could form a pit, pock, or hole. Blistering is mainly the result of volume expansion due to swelling, gas inclusion, gas formation, soluble impurities at the film/support interface from osmotic processes, or due to electroosmotic effects. Blistering is also an effect of hydrogen damage to alloys occurring predominantly in low-strength alloys when atomic hydrogen diffuses to internal defects and then precipitates as molecular hydrogen, which can locally plastically deform the alloy. Seen most often in exposures to aggressive corrosive environments (such as H_2S) or upon being pickled. *See also* **blister cracking**.

block coat—adhesive tie coat between incompatible paints. *Also known as* **transition primer**.

block copolymer—essentially a linear copolymer in which there are repeated sequences of polymeric segments of different chemical structure.

blooming—visible exudation or efflorescence on a surface. Also used to describe deposits resembling the bloom on a grape formed on paint films, which in most cases can be removed with a damp cloth. The term haze is used to indicate a permanent type of bloom. *See also* **blushing**.

blow-by—seepage of fuel or gases from the combustion chamber of an internal combustion engine into the crankcase as the result of high pressure differential, incomplete combustion, loose rings, etc.

blowdown—withdrawal of water from an evaporating water system to maintain a solids balance within specified limits of concentration of those solids. Also known as bleedoff.

blown—term applied to certain fatty oils that have been thickened by heating and blowing air through them. Also applied to bitumen whose melting point has been raised and physical properties modified by similar treatment.

blue brittleness—brittleness exhibited by some steels after heating to between 200 to 370 °C (400 to 700 °F), particularly if worked at this temperature.

blue scale—*see* **mill scale**.

blue vitriol—copper(II) sulfate pentahydrate ($CuSO_4 \cdot 5H_2O$).

bluing—formation of a blue-colored thin oxide film on steel by heating in air or by immersion in oxidizing solutions or salts.

blushing—whitening and loss of gloss of an organic coating caused by moisture. *Also called* **blooming**.

BOD—*abbreviation for* **biochemical oxygen demand**.

bodied oil—drying oil that has been heat polymerized.

body—loose term used to designate the relative viscosity or consistency of a paint, varnish, lacquer, or resin.

body fluids—human body fluids are distributed between intracellular (about 70%); interstitial (about 20%); and blood (10%). It is the interstitial fluid that is most potentially corrosive to metallic implants. This fluid is an aerated solution containing approximately 1% sodium chloride with minor amounts of other salts and organic compounds at 36 or 37 °C (97 to 99 °F).

boehmite (bohmite)—mineral form of mixed aluminum oxide and hydroxide ($AlO \cdot OH$).

Bohr model (of the atom)—states that in the atom the protons and neutrons are packed together in the center of the atom where they form the nucleus. Protons and neutrons have the same weight and are considered one unit, while electrons, which are almost weightless, travel around the nucleus in several orbits.

boiler deposits—accumulated material on boiler surfaces that can cause overheating and circulation restrictions. Results from the precipitation of soluble constituents in boiler feedwater. These commonly arise from soluble calcium, magnesium, iron, copper, and aluminum salts, from silica, and to a lesser extent from silt and oil.

boiler feedwater—water that directly enters a boiler. Usually consisting of makeup water and one or more sources of return condensate.

boiler fuels—liquid, solid, and gaseous fuels used to fire boilers. Includes natural gas, refinery gas, blast furnace gas, carbon monoxide, and other waste gases, coal, wood, solid wastes, waste oils, light oils, heavy oils, and other combustible liquids.

boilers—enclosed vessels in which water is heated and circulated, either as hot water or as steam, for heating or power. Generally classified as either firetube or watertube. In firetube boilers, hot combustion gases pass through tubes surrounded by water. In watertube boilers, water is converted to steam inside the tubes while hot gases pass on the outside.

boiler water—water used to produce steam.

boiler water terminology—definitions of the water in various parts of the boiler cycle: (1) condensate, which is steam that is condensed and returned to the boiler system, (2) makeup water, which is water added to the system to replace that lost due to process requirements, blowdown or leakage, and (3) feedwater, which is the condensate and makeup water returned to the boiler.

boiler water treatment—conditioning of water for boiler use to prevent scaling of boiler heat-transfer surfaces and eliminate corrosion so that the boiler operates efficiently and dependably. The conditioning is performed on the raw water supply, condensate return water, and the boiler water itself.

boiling out—cleansing of the internal surfaces and waterwalls of a new boiler of any foreign matter such as retained lubricants and temporary protective coatings. Usually conducted during fabrication or erection of the boiler.

boiling point—temperature at which the saturated vapor pressure of a liquid equals the external atmospheric pressure.

boiling-point elevation—increase in the boiling point of a liquid or solution after a nonvolatile solute is added. *See also* **colligative property.**

boiling-water reactor (BWR)—nuclear boiling-water reactor in which heat is transferred to the steam-raising boiler or heat exchanger by water used as both the coolant and moderator. *See also* **liquid-water reactor (LWR)** and **pressurized-water reactor (PWR).**

bond—the attachment at an interface between adhesive and adherend. Also the strong force of attraction holding atoms together in a molecule or crystal. *See also* **chemical bond.**

bond energy—amount of energy associated with a bond in a chemical compound.

Bonderize—proprietary term for processes for the phosphate conversion coating of iron or steel articles. *See also* **Parkerized.**

bonding electrons—electrons attracted by two nuclei; shared electrons.

bone black—charcoal made by heating bones and dissolving the calcium phosphates and other mineral salts with acid.

borate—any of a range of ionic compounds having negative ions containing boron and oxygen. Most borates are inorganic polymers with rings, chains, or other networks based on the planar BO_3 group or the tetrahedral $BO_3(OH)$ group.

borax—compound of the composition $Na_2B_4O_7 \cdot 10H_2O$.

boric acid—compound of the composition H_3BO_3.

boroscope—instrument used to visually inspect for boiler scale. A rigid instrument in which the image is brought from the objective to the eyepiece by an optical train consisting of an objective lens, prism, relay lenses, and an eyepiece lens.

bottom ash—solids falling to bottom of a boiler furnace. Also the larger, 100 microns or larger in diameter, slag particles (ash) removed at furnace bottoms of fossil fuel power plants.

boundary layer—thin layer of fluid around a solid body or surface relative to which the fluid is flowing.

boundary lubricant—either a lubricant in solid form or one formed on fresh metal surfaces by chemical reaction to provide for boundary lubrication.

boundary lubrication—lubricating contact condition where the pressures become too high or running speeds too low or surface roughness too great to prevent penetration of the lubricating film such that contact between asperities will occur.

Boyle-Charles law—the product of the pressure and volume of a gas is a constant that depends only upon the temperature. Applies only to an ideal gas.

Boyle's law—volume of a gas at a constant temperature is inversely proportional to its pressure. Applies only to an ideal gas.

brackish water—water containing low concentrations of any soluble salt. More specific definitions include (1) water having salinity values ranging from approximately 1000 to 30,000 mg/L, (2) water having less salt than seawater, but still undrinkable, and (3) EPA defines brackish water as water containing >10,000 mg/L dissolved salts.

branched carbon chain—chain of carbon atoms in which substituent hydrocarbon groups are bonded to interior carbon atoms.

brass—group of alloys made up of varying amounts of, principally, copper and zinc.

brazing—joining two pieces of metal together with a nonferrous filler metal at above 450 °C (840 °F) and below the solidus of the base metals.

breakdown potential—least noble potential where pitting, crevice corrosion, or both, will initiate and propagate.

breakdown voltage—voltage at which there is a sudden passage of a current through an insulator.

breakpoint chlorination—application of enough chlorine, after the chlorine demand has been satisfied, to maintain a free available chlorine residual.

breathing—permeability of a coating, e.g., latex coating that permits water vapor to pass through without detrimental effect on the film.

breath test—test for cleanliness of a metal surface. A clean surface produces a uniform clouding when breathed upon. A greasy one will show droplets.

brick—solid masonry unit, usually a rectangular prism, of clay or shale that has been burned or fired in a kiln. It is a ceramic product. Common bricks are made from clay by firing to about 950 to 1100 °C (1740 to 2010 °F).

brick masonry—one of oldest forms of building construction that utilizes bricks. Modern practice utilizes masonry with other structural materials, such as steel, concrete, and timber, to form an architectural cladding.

bright annealing—annealing metal in a protective medium to prevent discoloration of the bright surface.

bright blast—*see* **white blasting**.

bright dip—chemical solution used to produce a bright surface on a metal.

bright drawn steel—cold-drawn two-phase carbon steel that possesses a bright finish because atmospheric oxidation is eliminated and that has significantly increased mechanical properties due to the cold working.

brightener—addition agent that leads to the formation of a bright electroplate or that improves the brightness of an electroplated deposit.

brightening—chemical polishing of aluminum surfaces that is preserved by the subsequent protection given by anodizing. Most brightening employs phosphoric acid-nitric acid mixtures with other additions including hydrofluoric acid or ammonium bifluoride. *Also called* **chemical polishing**, bright dipping, and chemical brightening.

bright plating—process that produces an electrodeposit having a high degree of specular reflectance in the as-plated condition.

bright stock—cylinder oil further refined by solvent extraction or by acid and/or clay treatment.

brine—water containing dissolved matter at approximately more than 30,000 milligrams per liter, i.e., more dissolved salt than found in ocean water. In refrigeration terminology, a brine is any liquid cooled by a refrigerator and circulated as a heat-transfer fluid. Includes aqueous solutions of inorganic salts, e.g., sodium chloride and calcium chloride; aqueous solutions of organic compounds, e.g., alcohols and glycols; and chlorinated or fluorinated hydrocarbons and halocarbons, e.g., methylene chloride and trichloroethylene.

Brinell hardness—scale for measuring the hardness of metals.

brinelling—damage to a solid-bearing surface characterized by plastically formed indentations due to overloading.

British Standards Institution (BSI)—national organization (British) that establishes and publishes standard specifications and codes of practice.

British thermal unit (BTU or Btu)—quantity of energy required to raise the temperature of 1 pound mass of water 1 °F, averaged from 32 to 212 °F.

brittle fracture—separation of a solid accompanied by little or no plastic deformation. *Compare with* **ductile fracture**.

brittleness—relative degree of resistance to cracking or breaking by bending. The property of a material making it susceptible to the propagation of a fracture without appreciable deformation.

broaching—machining process in which a cutting tool having multiple transverse cutting edges is pushed or pulled through a hole or over a surface to remove metal by axial cutting.

bromic acid—compound of the composition $HBrO_3$. A strong acid.

bromine value—number of centigrams of bromine absorbed by 1 gram of test oil under certain conditions as a test for the degree of its unsaturatedness.

bronze—any of a group of alloys of principally copper and tin with lead and zinc sometimes present.

bronzing—application of a chemical finish to alter the surface color of copper or its alloys. Also the formation of a metallic sheen on a paint film.

Brownian motion (Brownian movement)—continual, irregular movement of extremely small particles suspended in a fluid, due, apparently, to their bombardment by molecules of the fluid.

brown ring test—test for ionic nitrates. Aqueous solution of a sample has added to it a solution of ferrous sulfate ($FeSO_4$). Concentrated sulfuric acid (H_2SO_4) is then added slowly so that it forms a separate layer. A brown ring of $Fe(NO)(SO_4)$ at the junction of the liquids shows presence of nitrate in the sample.

brown rot—internal attack of wood used in cooling towers by deterioration of its cellulose fibers due to thermophilic fungi such as of the Basidiomycetes group.

brucite—mineral form of magnesium hydroxide ($Mg(OH)_2$).

brush-off blast—lowest blast cleaning standard. This surface preparation standard is now designated "light blast cleaning" by the ISO. It is a surface that has its oil, grease, dirt, loose rust scale, loose mill scale, and loose paint removed.

brush plating—electrochemical process to selectively deposit metals onto a conductive surface by utilizing a hand-held anode conforming to the shape of the workpiece. Current passes through the plating solution that flows between the anode and workpiece. The anode is usually covered with an absorbent material to prevent its direct contact with the workpiece. Advantages of brush plating include its portability, good thickness control, low heat generation, and, usually, good adhesion of the plating. *Also known as* **selective plating**.

Brytal process—proprietary name for an electrolytic brightening process for polished aluminum effected by its anodic smoothing in an alkaline carbonate/phosphate electrolyte.

BSI—abbreviation for the British Standards Institution, which establishes and publishes standard specifications and codes of practice.

B-stage—intermediate stage in the reaction of certain thermosetting resins in which the material swells when in contact with certain liquids and softens when heated but may not entirely dissolve or fuse. *See also* **A-stage** and **C-stage**.

Btu (BTU)—British thermal unit. Amount of heat required to raise the temperature of one pound of water one degree Fahrenheit.

bubble—preferred term is blister.

buffer—compound or mixture that, when in solution, causes the solution to resist change in pH when an acid or alkali is added.

buffer solution—solution composed of a weak acid and its salt or a weak base and its salt and that resists changes in pH.

buffing—abrasive finishing operation on metals to produce smooth reflective surfaces by bringing the metal into direct contact with a revolving cloth or sisal buffing wheel charged with a suitable fine abrasive.

build—real or apparent thickness, fullness, or depth of a dried film.

bulk density—weight per unit volume of a material including voids inherent in the material as tested. Apparent mass per unit volume. *Also known as* **apparent density**.

burning—permanently damaging a metal or alloy by heating to cause either melting or intergranular oxidation.

burning off—removal of paint by subjecting it to heat and then scraping off while still soft.

burnishing—smoothing of surfaces by rubbing. Accomplished by movement rather than removal of the surface layer.

burn-off oven—oven used for heat paint stripping of hooks, racks, and hangers used in industrial painting to avoid water and waste pollution problems while improving coating operations. The components to be stripped are progressively heated from about 480 to 1380 °C (250 to 750 °F).

burnt deposit—rough, noncoherent, or otherwise unsatisfactory deposit from electroplating produced by the application of an excessive current density and usually containing oxides or other inclusions.

burnt lime—calcined limestone, either CaO or a mixed CaO·MgO. *Also called* **quicklime**.

burr—hanging sliver or a sharp edge on a material, usually metal, left by a previous manufacturing operation.

bus bar—rigid, conducting section for carrying current to electrodes (anode or cathode).

butadiene—organic compound of the composition $CH_2=CHCH=CH_2$. Widely used to synthesize various synthetic rubbers.

butyl rubber—synthetic rubber highly resistant to weather and heat and further characterized by low resiliency and low air permeability. Made by copolymerizing 2-methylpropene (isobutylene) and methylbuta-1-3-diene (isoprene).

BWR—*abbreviation for* **boiling-water reactor**.

C

CAA—*abbreviation for* **Clean Air Act**.

CaCO$_3$—chemical formula for calcium carbonate. Also used in the expression "CaCO$_3$ hardness," which is the concentration per unit volume of dissolved calcium and magnesium salts in a water where both are expressed as CaCO$_3$. *See also* **carbonate hardness**.

cadmium—metal having the nearest potential to that of aluminum in many environments, which accounts for the widespread use of cadmium-plated fasteners for direct contact with aluminum to reduce the potential for galvanic coupling. Except for low-melting alloys and some bearings, cadmium is used almost exclusively as an electroplated coating. Cadmium coatings are sometimes also used in place of zinc coatings on high-strength steels for their improved protection in moist atmospheres.

cadmium cell—*see* **Weston cell**.

cake—dewatered residue from a filter, centrifuge, or other dewatering device.

caking—settling of particles from a suspension into a compact mass that is not easily redispersed by stirring.

calcareous deposit—layer consisting of a mixture of calcium carbonate and magnesium hydroxide deposited on surfaces that are cathodically protected because of an increased pH adjacent to the protected surface.

calcimine—white or tinted liquid containing zinc oxide (ZnO), water, glue, and, optionally, coloring matter used as a "whitewash" for walls and ceilings.

calcination—process of heating or roasting a material to a high temperature, but below its fusing point, to cause it to lose moisture or other volatile material, or to be oxidized or reduced. Also less frequently used to describe the formation of a calcium carbonate deposit from hard water.

calcined gypsum—dry powder, primarily calcium sulfate hemihydrate (CaSO$_4 \cdot \frac{1}{2}$H$_2$O), from the calcination of gypsum. It is the cementatious base for most gypsum plasters. *Also called* **plaster of Paris** and gypsum cement.

calcite—naturally occurring form of calcium carbonate (CaCO$_3$). It is the essential material of limestone, marble, and chalk.

calcium carbonate saturation index—qualitative indication of the tendency of calcium carbonate in a cooling water to deposit or dissolve. Also known as the Langelier saturation index. It is the algebraic difference between the actual pH of a sample of water and its computed saturation pH. The latter is the pH at which water with a given calcium content and alkalinity is in equilibrium with calcium carbonate. *See also* **Langelier saturation index**.

calcium hardness—calcium concentration in water that is expressed as the concentration per unit volume, usually milligrams per liter, of calcium carbonate ($CaCO_3$).

calcium hydroxide—compound of the composition $Ca(OH)_2$. Used as an inexpensive alkali to neutralize acidity in soils, in the manufacture of mortar, and in making whitewash. *See also* **slaked lime**.

calcium oxide—compound of the composition CaO. Formed on the thermal decomposition of calcium carbonate ($CaCO_3$). Used to make calcium hydroxide. *See also* **quicklime**.

calcium plumbate—cream-colored corrosion-inhibitive paint pigment ($2CaO \cdot PbO_2$).

Calgon—proprietary name for a water-softening agent based on polyphosphates, which contain the ion $P_6O_{18}^{6-}$.

calomel half cell (calomel electrode)—electrode widely used as a reference electrode of known potential (–0.2415 volt at 25 °C) in the electrometric measurement of acidity, alkalinity, corrosion studies, and measurement of potential. Consists of a mercury electrode in contact with a solution of potassium chloride of specified concentration that is saturated with mercurous chloride (calomel).

calorie—quantity of heat that will raise the temperature of one gram of pure water by 1 °C, averaged from 0 to 100 °C.

calorizing—forming a protective coating on iron or steel articles by heating in aluminum powder at 800 to 1000 °C (1470 to 1830°F). Imparts improved resistance to oxidation. Also the trade name for a particular two-stage pack diffusion process for coating parts with a high concentration of aluminum.

capacitance—property of a system of conductors and dielectrics that permits the storage of electricity when potential difference exists between the conductors. Its value is expressed as the ratio of quantity of electricity to a potential difference. Its value is always positive.

capacitor—arrangement of conductors separated by an insulator (dielectric) that is used to store charge or introduce reactance into an alternating current circuit.

capillarity—force of attraction between similar and dissimilar substances.

capillary condensation—phenomenon whereby atmospheric water condenses into capillaries or crevices to produce a liquid environment in which there should be dryness for that relative humidity. It is brought about because the vapor pressure above a concave surface is less than the vapor pressure across a flat surface.

carbide—compound of carbon with one or more elements in which the carbon is considered less positive in character than the other elements.

carbohydrate—one of a group of organic compounds based on the general formula $C_x(H_2O)_y$. The simplest carbohydrates are known as the sugars. Carbohydrates of much greater molecular weight and structural complexity are known as polysaccharides, of which starch and cellulose are examples.

carbolic acid—*see* **phenol**.

carbon—element of the symbol C; a nonmetallic element in group IV of the periodic table. Has two main allotropic forms, diamond and graphite. The amorphous forms of carbon include forms such as carbon black and charcoal.

carbonaceous—matter containing carbon.

carbon adsorption—water remediation treatment in which contaminants are collected in condensed form on a carbon surface.

carbonate—any salt of carbonic acid containing the carbonate ion (CO_3^{2-}).

carbonate alkalinity—part of the total alkalinity due to the carbonate ion CO_3^{2-}).

carbonate hardness—that hardness in a water caused by bicarbonates and carbonates of calcium and magnesium. If alkalinity exceeds total hardness, all hardness is carbonate hardness; if hardness exceeds alkalinity, the carbonate hardness equals the alkalinity.

carbonation—process of chemical weathering in which minerals containing basic oxides are changed to carbonates by reaction with atmospheric carbon dioxide and water. With reference to the matrix of concrete, carbonation is the loss of its alkalinity from the ingress of acidic atmospheric oxides, e.g., carbon dioxide, sulfur dioxide, and sulfur trioxide. Carbonation also refers to the solution of carbon dioxide in a liquid under pressure. *See also* **anticarbonation coatings**.

carbon black—essentially elemental carbon as near-spherical colloidal particles obtained from the partial combustion or thermal decomposition of hydrocarbons.

carbon dioxide—CO_2; colorless, odorless, nontoxic gas under standard conditions, although high concentrations can cause stupefaction and suffocation because of displacement of ample oxygen for breathing. It is soluble in water to the extent of approximately 1 volume of CO_2 in 1 volume of water at 15 °C (59 °F) forming carbonic acid (H_2CO_3) in the process. Solid CO_2 sublimes at –79 °C (174 °F) at 760 torr. CO_2 is the principal corrosive agent in condensed steam generated in boilers by the thermal decomposition of carbonates and bicarbonates present in feedwater.

carbon dioxide blasting—blast cleaning technique that utilizes "dry ice" pellets (frozen carbon dioxide) in place of grit.

carbon dioxide corrosion—corrosion of the drill pipe in petroleum production due to formation carbon dioxide entering the drilling mud (or fluid). Carbon dioxide causes pitting primarily by under-deposit corrosion cell action. Prevention involves controlling pH in the higher alkaline regions, usually with calcium hydroxide, and using film-forming inhibitors of the oil-soluble amine type.

carbonic acid—compound of the composition H_2CO_3. Formed in solution when carbon dioxide is dissolved in water.

carbonitriding—case (surface) hardening process in which ferrous material is heated above the lower transformation temperature in a gaseous atmosphere of composition to cause simultaneous absorption of carbon and nitrogen by the surface and accompanying diffusion to create a concentration gradient. Usually involves heating the ferrous material in an atmosphere containing various combinations of hydrocarbons, ammonia, and carbon monoxide followed by a quenching to harden the case.

carbonize—to decompose an organic compound into carbon by heating.

carbon pickup—surface carburization of austenitic stainless steels during production due to casting into molds containing carbonaceous materials such as organic binders and washes or baked oil sand.

carbon potential—amount of carbon available for reaction in an environment and which amount depends upon the chemical balance of the carburizing and decarburizing agents in the system such as carbon monoxide, hydrogen, carbon dioxide, water vapor, methane, and nitrogen.

carbon steel—unalloyed steel. A steel containing carbon and additionally only minor amounts of other elements (less than 2%) whose presence is accidental and incidental and dependent upon the composition of the raw materials and the manufacturing process. *See also* **mild steel**.

carbonyl group—bivalent radical –CO– as it occurs in aldehydes and ketones.

Carborundum—trademark for a silicon carbide abrasive. *See also* **silicon carbide**.

carboxylate anion—anion that results when a carboxylic acid loses a H+. The general formula is *R*COO–.

carboxyl group—organic group –COOH present in organic acids.

carburization—absorption of carbon atoms into a metal surface at high temperatures.

carburizing—process in which an austenitized ferrous metal is contacted with a carbonaceous atmosphere, creating a concentration gradient of carbide formation.

carcinogen—agent producing or inciting cancerous growth in an organism.

cariogenesis—dental term corresponding to the ability for released metallic ions and formed corrosion products to affect the resistance of either dentin or enamel to decay (caries).

carrier fluid—fluid medium that transports impinging solid or liquid particles and that gives the particles their momentum relative to the solid surface.

carrier gas—gas that carries another gas, e.g., the sample gas in gas chromatography.

carryover—boiler water in steam caused by foaming or entrainment. Also used to describe any contaminant that leaves the boiler steam drum with the steam. It can be solid, liquid, or vaporous carryover.

cartridge filter—mechanical dust collector that uses pleated paper or polyester filter cartridges to clean dust-laden air.

CAS—abbreviation for the American Chemical Society's Chemical Abstracts Service Registry Number. It is a numeric designation uniquely identifying a specific chemical compound.

case—outer portion of ferrous alloy that has been made harder than the inner portion as a result of altered composition or structure from treatments such as carburizing, nitriding, or induction hardening.

case hardening—generic term covering several processes applicable to steel that change the chemical composition of the surface layer by absorption of carbon, nitrogen, or both, and which, by diffusion, create a concentration gradient. The outer portion, or case, is made substantially harder than the inner portion, or core. *See also* **cementation**.

casein—any of a group of phosphate-containing proteins found in milk.

casein paints—paints for masonry and plaster surfaces that contain about 10% casein with some lime to insolubilize the casein after application.

CASS test—*abbreviation for* **copper-accelerated salt spray test**. An accelerated corrosion test for some electrodeposits and for anodic coatings on aluminum. See ASTM B 368 for details.

cast iron—generic term for alloys consisting primarily of iron, and containing carbon and silicon where the carbon is in excess of that retained in solid solution in austenite at the eutectic temperature.

catabolism—metabolic process in which substances are degraded.

catalyst—substance used to change the rate of a chemical reaction without itself undergoing any permanent chemical change.

cataphoresis—*see* **electrophoresis**.

catastrophic oxidation—refers to metal-oxygen reactions that occur at continuously increasing and damaging rates. Also refers to oxidation that causes a protective metal oxide to break away from its underlying metal, thus losing protective behavior. *See also* **hot ash corrosion**.

catastrophic wear—rapidly occurring or accelerating surface damage caused by wear.

cathode—electrode where reduction is the principal reaction. A negative electrode.

cathode efficiency—current efficiency of a specified cathodic process.

cathode film—layer of solution in contact with the cathode that differs in composition from that of the bulk of the solution.

cathode scale—deposition of compounds from water and soils as a scale on cathodic sites as a result of the corrosion process.

cathodic cleaning—electrolytic cleaning in which the workpiece is made the cathode. *Contrast with* **anodic cleaning**.

cathodic corrosion—unusual condition in which metal loss is accelerated at the cathode because the alkaline condition there is corrosive to certain, primarily amphoteric, metals, e.g., aluminum, zinc, lead.

cathodic delamination—debonding of adherends (paints, rubber linings, etc.), development of leak paths under seals, or the blistering and peeling of coatings due to the action of a cathodic potential that is used otherwise for metallic cathodic protection.

cathodic disbondment—destruction of adhesion between a coating and its substrate by products of a cathodic reaction.

cathodic efficiency—current efficiency at the cathode.

cathodic inhibitor—substance or combination of substances that prevent or reduce the rate of the cathodic or reduction reaction by a physical, physicochemical, or chemical reaction.

cathodic pickling—electrolytic pickling in which the workpiece is made the cathode.

cathodic poison—substance that interferes with the cathodic reduction reaction. The rate of the cathodic reaction is slowed, and because anodic and cathodic reactions must proceed at the same rate, the whole corrosion process is slowed. Examples of cathodic poisons are sulfides and selenides. A drawback to using cathodic poisons as inhibitors is that they sometimes cause hydrogen blistering of steel and increase its susceptibility to hydrogen embrittlement.

cathodic polarization—change of the electrode potential in the active (negative) direction due to current flow.

cathodic protection (CP)—reducing corrosion by making the metallic object a cathode of an electrochemical cell using impressed direct current or by its attachment to a sacrificial metal anode, e.g., magnesium, zinc, aluminum. Cathodic protection is used in many applications such as with steel piles, well casings, interior components of structures such as pumps, with condenser water boxes, pipelines, storage tanks, buried structures including gas and oil pipelines, pier supports, offshore drilling platforms, hulls of ships, ship propellers, and steel bulkheads. *Also known as* **sacrifical protection**.

cathodic reaction—electrode reaction equivalent to a transfer of negative charge from the electronic to the ionic conductor. It is a reduction process.

cathodic sputtering—method of forming a coating of a corrosion-resistant metal on a corrosion-prone substrate metal by discharging an inert gas atmosphere such that the positively charged gas ions are attracted to the cathode where they dislodge atoms from the resident corrosion-resistant species which, in turn, are attracted to the corrosion-prone anode metal and coat it.

catholyte—electrolyte adjacent to the cathode of an electrolytic cell. *Contrasts with* **anolyte**.

cation—positively charged ion.

cation exchange material—material capable of the reversible exchange of positively charged ions.

cationic detergent—detergent that produces positively charged colloidal ions in solution.

cationic emulsion—emulsion in which the emulsifying system establishes a predominance of positive charge on the discontinuous phase.

caulk—to fill voids with plastic or semiplastic material.

caustic—substance that is strongly alkaline. Also a hydroxide of a light metal, such as sodium hydroxide or potassium hydroxide. Also called caustic corrosion.

caustic attack—boiler corrosion resulting from the presence of excessive sodium hydroxide. Also called caustic corrosion.

caustic corrosion—*see* **caustic attack**.

caustic cracking—stress corrosion cracking of metals in caustic solutions.

caustic dip—strongly alkaline solution into which metal is immersed for etching, for neutralizing residual etching, or for removing greases or paints.

caustic embrittlement (intercrystalline cracking)—cracking of metals in caustic solutions from the combined actions of tensile stress and corrosion.

caustic potash—potassium hydroxide (KOH).

caustic soda—sodium hydroxide (lye) (NaOH).

cavitation—formation and sudden collapse of vapor or gas bubbles within a liquid in motion when the pressure is reduced to a critical value while the ambient temperature remains constant.

cavitation cloud—collection of a large number of cavitation bubbles.

cavitation corrosion—process involving conjoint corrosion and cavitation.

cavitation damage—damage to a surface caused by severe turbulent flow resulting in cavitation.

cavitation erosion-corrosion—corrosion damage from the conjoint action of cavitation, erosion, and corrosion.

cavitation erosion damage—progressive loss of original material from a solid surface due to the conjoint action of cavitation and erosion.

CCT—*abbreviation for* **cyclic corrosion testing**.

CDA—abbreviation for the Copper Development Association.

cedar—durable softwood noted for its decay resistance.

cell—electrochemical system consisting of an anode and cathode immersed in an electrolyte. In an electrolytic cell, current from an outside source passes through the electrolyte to produce chemical change. In a voltaic cell, spontaneous reactions between the electrodes and electrolyte produce a potential difference between the two electrodes.

cellophane—thin, flexible, transparent cellulosic material regenerated from viscose and containing variable amounts of water and softener and coated to render itself as a moisture-proof wrapping.

cellular material—generic term for materials containing many cells (open, closed, or both) dispersed throughout the mass.

celluloid—transparent, highly flammable material made from cellulose nitrate using a camphor plasticizer.

cellulose—carbohydrate (polysaccharide) that is the principal constituent of wood.

Celsius—designation of the degree on the International Temperature Scale. Formerly called centigrade. Symbol is C.

cement—any of various substances used for bonding or setting to a hard material. Portland cement is a mixture of calcium silicates and aluminates made by heating limestone with clay (containing aluminosilicates) in a kiln. The product, ground to a fine powder and mixed with water, sets in a few hours then hardens over a longer period due to the formation of hydrated aluminates and silicates.

cement-asbestos board—rigid, noncombustible board highly resistant to weathering that consists of asbestos fibers bonded with Portland cement.

cementation—any metallurgical process in which the surface of a metal is impregnated by some other substance. *See also* **case hardening**. Also used to describe metal ion deposition. *See also* **metal ion deposition**.

cementatious materials—include Portland and blended hydraulic cements, fly ash and other pozzolans, and ground granulated blast-furnace slag.

cementite—compound of iron and carbon known chemically as iron carbide and having the approximate formula Fe_3C.

cementitious surfaces—concrete surfaces generally classified by type as either: poured or precast slab, block wall, gunned or troweled surfaces, or flooring when being considered for their surface preparation and coating.

cement paint—paint supplied in dry powder form composed primarily of Portland cement and colorant pigments. It is mixed with water immediately before use.

centigrade—obsolete designation of the degree on the International Temperature Scale. *See* **Celsius**.

centipoise—standard unit of viscosity equal to 0.01 poise, the c.g.s. unit of viscosity.

centrate—liquid remaining after removal of solids by centrifugation.

centrifugal blast cleaning—process using motor-driven bladed wheels to hurl abrasives by centrifugal force to clean a surface.

ceramic—any of various hard, brittle, heat-resistant, and corrosion-resistant materials made by firing (heat treating) clay, other minerals, or synthetic inorganic compositions. Consists of one or more metals in combination with a nonmetal, usually oxygen.

ceramic coatings—includes porcelains and high-temperature coatings based on oxides, carbides, silicides, borides, nitrides, cermets, or other inorganic materials.

ceramic glass enamel—vitreous inorganic coating bonded to glass by fusion at a temperature generally above 500 °C (932 °F).

cermet—composite material consisting of a ceramic in combination with a sintered metal.

CFC—*abbreviation for* **chlorofluorocarbon**.

CFR—abbreviation for the U.S. Code of Federal Regulations.

c.g.s. system—abbreviation for the centigrade-gram-second system for describing physical quantities. Now replaced by the SI units.

chalcogen—name applied to the group VIA elements, i.e., oxygen, sulfur, selenium, tellurium, and polonium.

chalk—fine-grained white rock composed of the skeletal remains of microscopic sea creatures and consisting largely of calcium carbonate ($CaCO_3$).

chalking—degradation of a paint film by the erosion of its binder due to the action of ultraviolet light, moisture, oxygen, and chemicals resulting in a loose, chalky, removable powder on the surface.

chalky—having a matte surface but with a powdery surface layer that can be wiped off with a finger touch.

char—carbonaceous material formed by pyrolysis or incomplete combustion of a combustible material.

characteristic X-ray analysis—analysis of a sample, usually a microminiature sample, to identify its elemental constituents using the scanning electron microscope.

charcoal—porous form of carbon produced by the destructive distillation of organic material.

charge (electric charge)—property of some elementary particles that gives rise to an interaction between them and consequently to the host of material phenomena described as electrical. Charge occurs in nature in two forms conventionally described as positive and negative in order to distinguish between the two kinds of interaction between particles.

charge carrier—entity that transports electric charge in an electric current. The nature of the carrier depends on the type of conductor: in metals, the charge carriers are electrons; in semiconductors, the carriers are electrons (*n*-type) or positive holes (*p*-type); in gases, the carriers are positive ions and electrons; in electrolytes, they are positive and negative ions.

charge density—electric charge per unit volume of a medium or body; or the electric charge per unit surface area of a body.

charge transfer—electrochemical process in which there is a transfer of a metal atom from its site in a metal lattice to an aqueous solution as the cation M^{n+} or as a hydrolyzed or complexed species (the anodic charge transfer) and that is accompanied by a reduction reaction at the cathode.

charge transfer overpotential—electrode overpotential resulting from the resistance offered to the transfer of electric charges through the double layer.

Charles' law (Gay-Lussac's law)—volumes assumed by a given mass of a gas at different temperatures, the pressure remaining constant, are directly proportional to the corresponding absolute temperatures.

checking—development of slight breaks in a coating film that do not penetrate to the underlying surface. Also cracking in a cross-hatch manner resembling mud cracking. Usually forms as the coating ages and becomes harder and more brittle.

chelant corrosion—dissolving or thinning attack of boiler tubes brought on by using excessive concentrations of sodium salt chelants for deposit solubility control over a period of many months. The attack concentrates on areas of stress within the boiler. Annealed tubes are not normally attacked.

chelate—(1) molecular structure in which a heterocyclic ring can be formed by the unshared electrons of neighboring atoms. (2) coordination compound in which a heterocyclic ring is formed by a metal bound to two atoms of the associated ligand.

chelating agent—complexing agent that, in aqueous solution, renders a metallic ion inactive through the formation of a chelate compound. *Also known as* **sequesterant**. *See also* **complexation**, **ligand**, and **coordination compound**.

chelation—formation of a soluble metallic complex. *See also* **sequester**.

chemical—of or pertaining to chemistry. Also of or pertaining to the properties or actions of chemicals. Also a substance produced by or used in a chemical process.

chemical bond—strong force of attraction holding atoms together in a molecule or crystal. Chemical bonds are of various types; ionic or electrovalent bonds, covalent bonds, coordinate bonds, or coordinate covalent bonds. *See also* **bond**.

chemical cell (battery)—cell in which the emf is produced by a chemical difference, e.g., a zinc electrode immersed in a solution of zinc sulfate (a half cell), which is connected by a salt bridge (connecting electrolyte) to a half cell of a copper electrode immersed in a solution of copper sulfate for a positive emf and supply of current at the electrodes.

chemical change—change by which the chemical composition of a substance is altered.

chemical coating—may refer to a thick coating of plastic or rubber, cement, glass, or ceramic, but most often is used to refer to the coating produced by chemical or electrochemical reaction with a metal surface. *See also* **conversion coating** and **anodizing**.

chemical conversion coating—protective or decorative nonmetallic coating produced on the surface of a metal by its reaction with a chosen environment. *Also called* **chemical coating**, chemical reaction coating, or **conversion coating**.

chemical corrosion—corrosion that takes place under the influence of dry gases (high-temperature corrosion) or in water-free organics and is not electrochemical in nature.

chemical engineering—study of the design, manufacture, and operation of plant and machinery in industrial chemical operations.

chemical equation—way of denoting a chemical reaction using specified symbols for the participating particles (atoms, molecules, ions, etc.).

chemical equilibrium—closed chemical system in which the forward and reverse reaction rates are equal.

chemical etching—*see* **chemical milling**.

chemical fluids—*see* **synthetic fluids**.

chemical formula—combination of chemical symbols with appropriate subscripts that indicates the ratio of atoms within molecules and formula units.

chemical lead—lead alloy containing about 0.06% copper and used for acid corrosion protection, particularly where sulfuric acid below 70% concentration is encountered and steel is attacked. More resistant than high-purity lead. *See also* **hard lead**.

chemically induced corrosion—term applicable to the rapid corrosion of boiler tubes caused by overheating where a heat-induced concentration gradient causes higher concentration of the corrodent at the tube surface.

chemical milling—process whereby a thin surface film of metal is removed from a metal surface by the corrosive action of an acid or alkali. Also known as chemical etching.

chemical oxygen demand (COD)—measure of organic matter and other reducing substances in water. It is the quantity of oxygen required for their oxidation and is obtained under specified standard test conditions.

chemical polishing—procedure for providing a smooth and bright surface by immersion in an aqueous acidic bath that also contains special surfactants and stabilizers for improving brightness and extending bath life. Primarily used on stainless steel and aluminum. Acids utilized include phosphoric, nitric, sulfuric, and hydrochloric for stainless steel, and sulfuric, nitric, phosphoric, acetic, chromic, and hydrofluoric for aluminum. *Also called* **brightening**.

chemical potential—free energy per mole of a substance. It is the change in Gibbs free energy with respect to change in amount of the component, with pressure, temperature, and amounts of other components being constant. Components are in equilibrium if their chemical potentials are equal. For ions, the chemical and electrical potentials have to be considered separately; *see* **electrochemical potential**.

chemical property—property that describes a chemical change that a substance undergoes.

chemical reaction—change in which one or more chemical elements or compounds (reactants) form new compounds (products). All reactions are to some extent reversible, i.e., the products can also react to give the original reactants. However, in many cases the extent of this back reaction is negligibly small and the reaction is regarded as irreversible.

chemical-reaction coating—*see* **conversion coating**.

chemical resistance—ability to resist chemical attack.

chemical symbol—either the one or two (or, in the case of newly discovered elements, three) letters used to represent each element. Usually, these letters are derived either from the modern or classical name of the element.

chemical vapor deposition—coating process similar to gas carburizing and carbonitriding whereby a reactant atmosphere gas decomposes at the surface of the work metal liberating a substance for either absorption or accumulation on the work metal. A second substance is liberated in gas form during the process and is removed from the processing chamber along with excess atmosphere or carrier gas.

chemisorption—adsorption followed by a chemical reaction so that the binding force of the adsorbate is at least partially chemical. *Contrast with* **physisorption**.

chemistry—study of matter and its interactions, particularly the elements and the compounds they form.

chilled iron—cast iron caused to solidify as white cast iron, locally or entirely, by accelerated cooling.

china—glazed or unglazed vitreous ceramic whiteware made by the china process that refers to firing a ceramic body to maturity followed by its glazing then maturing this by firing at a lower temperature.

chipping—cleaning of steel surfaces using special hammers.

chiral carbon atom—carbon atom that is bonded to four different groups.

chloride-induced stress-corrosion cracking—specific type of stress-corrosion cracking induced by a chloride concentration cell. Occurs when tensile stress accompanies the presence of solution chloride.

chlorinated hydrocarbon—organic compound having hydrogen atoms and chlorine atoms in its chemical structure, e.g., trichloroethylene, methylene chloride.

chlorinated rubber—resin formed by the reaction of rubber with chlorine or by chlorinating synthetic polymers such as 1-butene or polyethylene. Unlike rubber, the product is readily soluble in hydrocarbon solvents. The coatings from these usually have outstanding chemical resistance.

chlorination—addition of chlorine or chlorine-releasing agent to a water for purposes of disinfection. Prechlorination is the application of chlorine to water prior to other treatment processes. Rechlorination is the application of chlorine, at one or more points of a system, to water that has already been chlorinated. Postchlorination is the application of chlorine to a water effluent from other treatment processes. Dechlorination partially or completely reduces residual chlorine by chemical or physical means. Also a chemical reaction in which chlorine is introduced into a compound.

chlorine—Cl_2; an oxidizing, gaseous toxicant used as a disinfectant for water supplies and for the removal of taste and color from water. On contact with water, chlorine forms two acids, hypochlorous (HOCl) and hydrochloric (HCl) with the concentration of the former primarily determining the biocidal activity.

chlorine demand—usually expressed in parts per million or milligrams per liter, it is the difference between the amount of chlorine applied and the amount of free, combined, or total available chlorine remaining at the end of the contact period.

chlorine dioxide—ClO_2; an oxidizing, gaseous toxicant used as a disinfectant for water supplies. Unlike chlorine, which is used for the same purpose, chlorine dioxide does not form hypochlorous and hydrochloric acids in water, but exists solely as dissolved chlorine dioxide.

chlorine donors—chemicals that release active chlorine in water and, therefore, whose action is similar to chlorine, i.e., an oxidizing biocide. Commonly used chlorine donors include sodium dichloro-*s*-triazine trione or sodium dichloroisocyanurate and 1,3-dichloro-5,5 dimethylhydantoin.

chlorine, free available—hypochlorite ions (OCl^-), hypochlorous acid ($HOCl$) or the combination present in water.

chlorine requirement—amount of chlorine required to achieve, under specified conditions, the objectives of chlorination.

chlorine residual—amount of available chlorine present in water at any specified time.

chlorinity—measure of the chloride content, by mass, of seawater (grams per kilogram). Now defined as 0.3285233 times the weight of silver equivalent to all the halides present.

chlorofluorocarbon (CFC)—compounds in which some or all of the hydrogen atoms of a hydrocarbon have been replaced by chlorine and fluorine atoms. Previously widely used as aerosol propellants, refrigerants, in packaging, and as solvents. Because of their reactivity with ozone in the upper atmosphere, their use is greatly diminished.

chlorosity—concentration of dissolved chloride equivalent in water at 20 °C (68 °F).

chromadizing—improving paint adhesion on aluminum by treatment of the aluminum with a solution of chromic acid. *Also called* chromodizing, **chromating**, and chromatizing.

chromate conversion coating—chromate coating, on metallic surface produced by chemically converting the surface with an aqueous solution of chromates, or dichromates, and certain other organic and inorganic chemicals as additives. *See also* **chromating** and **conversion coating**.

chromate rinse—process for rinsing an iron phosphate or zinc phosphate conversion coated metal in an aqueous chromate solution to passivate it.

chromate treatment—process of treating a metal with a solution of a hexavalent chromium compound to produce a conversion coating.

chromating—process for producing a conversion coating containing chromium compounds on metal surfaces. Metals usually chromated include aluminum, cadmium, copper, magnesium, silver, zinc, and their alloys. *Also called* **chromadizing**.

chromatography—technique for analyzing or separating mixtures of gases, liquids, or dissolved substances based on the principle that different components of a sample are absorbed to different extents and move on the absorbent at different rates. All types of chromatography involve two stages, a stationary phase, which is the absorbent material, and a moving phase, which is the gas, liquid, or dissolved substances.

chrome lignite—mined lignite to which chromate has been added and reacted. Used as thinner and/or emulsifier for drilling muds.

chrome pickle—producing a chromate conversion coating on magnesium for its temporary protection or to serve as a paint base.

chromic acid—hypothetical acid H_2CrO_4. Term is usually used to describe an acidic solution of a dichromate. CrO_3, the anhydride of chromic and dichromic acids, is also called chromic acid.

chromium steel—any of a group of stainless steels containing about 8 to 25% chromium.

chromizing—surface treatment at elevated temperature, generally carried out in a pack, vapor, or salt bath, in which an alloy is formed by the inward diffusion of chromium into the base metal.

chronic—being of long duration, continuing, or constant.

chronic toxicity—substance property that causes adverse effects in an organism upon repeated or continuous exposure over a period of at least one-half the lifetime of that organism.

CIE—abbreviation for Commission Internationale d'Éclairage, the French name for the International Commission on Illumination.

cinder block—lightweight masonry unit made of cinder concrete and widely used for interior and exterior partitions.

circulating rate—cooling water term describing the amount of water cycled through the system in unit time, usually expressed in gallons per minute.

cladding—forming of a thin coating of one metal on another metal by arc or gas welding, roll bonding, explosion bonding, melting or brazing, soldering, or by some other physical means.

clad metal—composite metal containing two or more layers that have been bonded together, usually by corolling, coextrusion, welding, diffusion bonding, casting, chemical deposition, or electroplating.

clad steel—composite plate made of steel with a cladding of corrosion-resistant or heat-resistant metal on one or both sides.

clapboard—wood siding used as an exterior covering on buildings and that is applied horizontally and overlapped.

clamshell marks—*see* **beach marks**.

clarification—coagulation, flocculation, and sedimentation processes applied to waters for removal of their suspended solids, finer solids appearing as turbidity and color, and other colloidal materials.

clarifying agent—substance used in the clarification purification of various solutions and liquors.

Clark cell—voltaic cell consisting of a zinc amalgam anode and mercury cathode, both immersed in a saturated solution of zinc sulfate. With an emf of 1.4345 volts at 15 °C (59 °F), it has been used as a standard cell.

Clark process (clarking)—water softening process in which temporary hardness, but not permanent hardness, is removed from water by adding lime (calcium hydroxide) to the water, which causes the soluble calcium hydrogen-carbonate to precipitate as calcium carbonate.

clay—generic name for natural mineral aggregate consisting essentially of hydrous aluminum silicates, formed by the decomposition of feldspar and other aluminum silicates.

clean—free of contaminants.

Clean Air Act—federal act passed to protect against the accidental industrial release of hazardous vapors by designing and maintaining a safe plant, identifying hazards, minimizing consequences of accidental releases, and developing risk management plans if stated substances exceed threshold quantities.

clean air standards—set of enforceable rules, regulations, standards, limitations, orders, controls, prohibitions, etc., that are contained in, issued under, or adopted pursuant to the Clean Air Act.

cleaning—process of removing soil or unwanted material from a surface, generally by one or more of the following processes: detergency (the lifting of soil by displacing it with surface active materials), mechanical removal (wiping, brushing, or abrasion), solubilization (use of water, petroleum fraction, alcohol, or chlorinated hydrocarbon to dissolve the soil), and chemical reaction to yield soluble or noninterfering products.

cleanliness test—determination of the degree of cleanliness of a surface. Common cleanliness tests for metal surfaces include: the visual test (for spots and films); wiping test (with white cloth or tissue to pick up and view particulate matter and smut); water-break test (capacity of the hydrophilic metal surface to hold a complete film of water without showing any breaks in the film); atomizer test (spraying the dried metal surface with a mist of water containing dye to identify any retained discrete droplets due to presence of soil); fluorescence test (rinsing with an aqueous fluorescent dye solution and viewing with UV light for fluorescing soil); contact angle test (measuring cleanliness by the contact angle of a droplet of water); weight of residual soil test (gravimetric determination); and spray pattern or mist test (spraying a previously acid-rinsed metal surface with mist of water to identify any retained discrete droplets due to soil).

Clean Water Act of 1977 (CWA)—gives the Environmental Protection Agency (EPA) authority to regulate and control pollutants entering rivers, streams, lakes, and other waterways by their direct as well as indirect discharge.

clear—complete lack of any visible nonuniformity when viewed in mass by strong transmitted light.

clearcoat—transparent, nonpigmented organic coating applied for protection, aesthetic appeal, chemical resistance, gloss, or texture control.

Cleveland Condensing Humidity Cabinet (QCT)—accelerated weathering apparatus that utilizes condensation type of water exposure at elevated temperature.

climate—pattern of weather elements (temperature, rainfall, humidity, solar insolation, wind, etc.) in a given area over a time period.

climate cabinet—any enclosure used in simulating selected climatic conditions.

CLIMAT test—abbreviation for the classification of industrial and marine atmospheres test. An atmospheric galvanic corrosion test for metals that is also called the "wire-on-bolt" test. In this test, a wire of the anodic material is wrapped around a threaded rod of the cathodic material and atmospherically exposed for a short time. Subsequent examination is made for corrosion weight loss and/or signs of pitting.

closed recirculating cooling water system—recirculating cooling water system characterized by its essentially permanent charge of water, i.e., little or no losses due to evaporation, drift, blowdown, etc.

CMC—abbreviation for carboxymethyl cellulose, usually employed as water-soluble sodium carboxymethylcellulose, a common thickening and suspending agent additive.

coacervation—collection of emulsoid particles into liquid droplets preceding flocculation and which condition is reversible. It is an intermediate stage between sol and gel formation.

coagulation—process by which colloidal particles come together to form larger masses. Also process whereby a liquid is changed into a thickened, curdled, or congealed mass. Also the result of neutralization of charges on colloidal matter. Sometimes described by the term flocculation.

coagulum—agglomerate of particles.

coal—dark brown to black carbonaceous deposit derived from the accumulation and alteration of ancient vegetation.

coal conversion—conversion of coal by gasification to fuel gas and its hydrogenation to produce either liquid fuel (oil) or a clean solid fuel.

coalescence—fusing or flowing together of liquid particles into a continuous phase. Also the action of joining of particles into a film as the volatile components of a coating evaporate.

coal tar—dark brown to black cementatious material produced by the destructive distillation of coal.

coal tar epoxy paint—paint wherein the binder or vehicle is a combination of coal tar and epoxy resin.

coal tar urethane paint—paint wherein the binder or vehicle is a combination of coal tar with polyurethane resin.

coastal oil—naphthenic-type petroleum oil refined from Gulf and Pacific coast crudes.

coat—to cover a substrate surface with a layer of material, or the cover so applied.

coating—generic term for paints, lacquer, enamels, etc., applied as a liquid, liquefiable, or mastic composition, which is then converted to a solid or semisolid protective, decorative, or functional adherent film.

coating system—term encompassing the applied and cured multilayer paint film.

coat of paint—one layer of dry paint resulting from a single wet application.

COD—*abbreviation for* **chemical oxygen demand**.

coefficient of friction—ratio of the force resisting tangential motion between two bodies to the normal force pressing these bodies together.

coefficient of variation—*see under* **statistical terms**.

cohesion—mutual attraction by which the elements of a body are held together. Also the propensity of a single substance to adhere to itself. *Compare with* **adhesion**.

coil coating—process by which a continuous coil of metal is unwound, cleaned, surface-treated, coated, heat-cured, cooled, and rewound in one operation. The metal is postfabricated later to desired products. *Also called* **strip coating**.

coinage metals—group of three malleable ductile metals, copper, silver, and gold, that form subgroup IB of the periodic table.

coke—form of carbon made by the destructive distillation of coal.

coke laydown (coking)—deposition of carbon during cracking operations in the petrochemical industry that may result in the plugging of furnace tubes.

cold cracking—weld cracking, usually occurring below 205 °C (400 °F) during or after cooling to room temperature.

cold forming—term embracing such technology as cold extrusion, cold heading, wire and bar drawing, tube drawing, and deep drawing.

cold heading—high-speed method of gathering metal in certain sections along lengths of bar, rod, or wire by causing the metal to flow between the dies without preheating the metal. Used to make fasteners of various kinds.

cold lime softening—precipitation softening process for water accomplished at ambient temperatures. *See also* **warm lime softening**.

cold-rolled steel—low-carbon, cold-reduced, and annealed sheet steel.

cold stripping—method of paint stripping primarily based on immersion in a solution or bath of these key ingredients: methylene chloride, phenolics, and either alkaline and/or acid activators.

cold water pitting—electrochemical pitting of copper tubes and fittings in domestic water systems that is generally associated with ground waters containing free carbon dioxide in conjunction with dissolved oxygen and that may be accelerated by chlorides and sulfates in the water.

cold working—deforming metal plastically under conditions of temperature and strain rate that induce strain hardening. Cold working usually involves rolling or forging of the metal and usually, but not necessarily, conducted at room temperature. *Contrast with* **hot working**.

colligative property—property of a solution that depends on the number of dissolved nonvolatile solute particles rather than their type. *Examples include* **freezing-point depression, boiling-point elevation**, and vapor pressure lowering.

colloid—matter of very fine particle size, usually 10^{-5} to 10^{-7} centimeter in diameter. Colloidal particles tend not to settle when free standing. Colloids are classified as either sols that are dispersions of solid colloidal particles in a liquid, emulsoids in which both the dispersed and continuous phases are liquids, or gels where both dispersed and continuous phases have a three-dimensional network throughout the material that forms a jellylike mass.

colloidal solution—solution containing colloidal particles. *Sometimes called* **sol**.

collodion—thin film of cellulose nitrate. Usually made by dissolving cellulose nitrate in ethanol or ethoxyethane, coating a surface, and then allowing the solvent to evaporate.

color anodizing—term used in anodizing aluminum. The formation of a colored coating on aluminum where the colored compound, as pigment or dye, is incorporated after the anodized film has been formed.

colorant—any substance that imparts color to another material or mixture and may be either a dye or pigment.

color end point—that point during a titration when the solution color changes distinctly.

colorimetric analysis—quantitative analysis of solutions by estimating their color and comparing it with the color of standard solutions.

coloring—production of desired colors on metal surfaces by appropriate chemical or electrochemical action.

columbium—former name for the element niobium (Nb).

combination reaction—reaction in which two or more reactants are chemically combined to produce one product.

combined carbon—part of the total carbon in steel or cast iron present as other than free carbon.

combined water—water in a substance that is chemically held as water of crystallization.

combining weight—equal to the atomic weight of an element or radical divided by its valence.

combustible liquid—liquid having a flashpoint at or above 37.8 °C (100 °F).

combustion—chemical process of oxidation occurring at a rate fast enough to produce temperature rise and light either as a glow or flame.

commercial blast—moderate grade of blast cleaning. Now designated as thorough blast cleaning by the ISO. *See also* **white blasting**.

comminution—process by which agglomerates are reduced in size.

common ion effect—reversal of ionization that occurs when a compound is added to a solution of a second compound with which it has a common ion, the volume being kept constant.

compatibility—ability of two or more substances mixed with each other to form a stable homogeneous mixture or solution.

completion fluids—oil-field term describing the clear completion (and workover) fluids used in drilling operations where if a drilling mud fluid were used the drilling mud solids might cause formation damage to the oil sands by invading their porosity and severely reducing oil flow. Completion fluids are usually clear solutions of either sodium, potassium, or calcium chloride of increasing solution density as needed, or are mixed solutions of calcium chloride and bromide, or are zinc bromide solutions.

complex—compound in which molecules or ions form coordinate bonds to a metal atom or ion.

complexation—formation of complex chemical species by the coordination of groups of atoms, termed ligands, to a central ion, commonly a metal ion. Generally, the ligand coordinates by providing a pair of electrons that form an ionic or covalent bond to the central ion. *See also* **chelate**, **coordination compound**, and **ligand**.

complexing agent—compound that will combine with metallic ions to form soluble complex ions.

complex compound—chemical compound made up structurally of two or more compounds or ions.

complex ion—ionic species of a coordination or complex compound. Ion composed of two or more ions or radicals, both of which are capable of independent existence.

complex silicates—dense, insulating-type deposits, usually silicate compositions of sodium and some metal such as iron and aluminum, which can cause boiler tube failure.

complex soap grease—grease in which the soap crystals or fibers are formed by the cocrystallization of two compounds, a normal soap and a complexing agent such as water, a salt, or other additive.

compliance—generally refers to meeting specified clean air or water standards. Also refers to meeting a schedule or plan ordered by a court, the EPA, or any air or water-pollution control agency.

compliance coating—coating system that meets all air, water, and waste-disposal regulations.

composite—homogeneous material produced by the synthetic assembly of two or more materials.

composite sample—combination of two or more samples.

composition—analysis; makeup of a substance.

compound—substance formed by the combination of elements in fixed proportions. Formation of a compound involves a chemical reaction, i.e., there is a change in the configuration of the valence electrons of the atoms. Compounds, unlike mixtures, usually cannot be separated by physical means.

compounded oil—petroleum oil that has added animal or vegetable oil or some other substances.

Comprehensive Environmental Response Compensation and Liability Act of 1980 (CERCLA)—better known as the "Super-Fund Law," authorizes funding for cleanup of hazardous waste and assigns liability for guilt.

compressive strength—maximum compressive stress a material is capable of sustaining without fracture or excessive flattening.

compressive stress—stress that causes an elastic body to deform in the direction of the applied load.

concentrated—solutions in which a relatively large quantity of solute is dissolved in a solvent.

concentration—process of increasing the dissolved solids per unit volume of solution. Also the amount of material dissolved in a unit volume of solution.

concentration cell—electrolytic cell whose emf is caused by a difference in concentration of some component in the electrolyte. Also a connection of two solutions of the same composition, but different concentrations, by a metal conductor to produce current flow through the circuit. *See also* **concentration potential**.

concentration polarization—polarization of an electrode caused by concentration changes in the environment adjacent to the metal surface.

concentration potential—potential step associated with any boundary separating differing concentrations of ions. For monovalent ions, the concentration potential is about 59 millivolts for each tenfold change in concentration. *See also* **concentration cell**.

concentration ratio—ratio of the concentrations of salts in circulating cooling water to those of makeup water used with the system.

concrete—generally homogeneous mixture of Portland cement, aggregates, and water, which cures or hardens on standing. Also any mixture of fine and coarse aggregates firmly bound into a monolithic mass by a cementing agent.

concrete block—hollow or solid concrete masonry unit consisting of Portland cement and suitable aggregates, and possibly other admixtures, combined with water. Sometimes called cement block.

concrete cancer—degradation of concrete parking lot decks and pillars due to alkali-silica attack. Alkaline constituents in aggregate can react forming a gel that swells by absorbing water. It is not the result of chloride ion attack.

condensate—water obtained by its evaporation and subsequent condensation.

condensate corrosion inhibitors—volatile neutralizing amines used to control corrosion in heat exchangers caused by carbon dioxide dissolved in steam condensate. The most commonly used condensate corrosion inhibitors include morpholine, cyclohexylamine, diethylaminoethanol (DEAE), and dimethylamino-2-propanol (DMA-2-P). In the case of brass condenser tubes, the tubes are susceptible to condensate corrosion in steam condensate containing high concentrations of ammonia and oxygen. This is sometimes called ammonia attack. This can generally be prevented by eliminating air from the steam and keeping the condenser tubes clean.

condensate polishing—use of demineralizer systems to purify condensate water used for boilers in order to maintain the purity required for operation of once-through boilers.

condensate wells—term applied to high-pressure oil and gas wells (up to 10,000 lb/in.^{2}) whose fluids are a gas containing dissolved hydrocarbons.

condensation—change of a vapor or gas into a liquid. Also a chemical reaction where two or more molecules combine with the separation of water or some other simple substance.

condensation exposure—exposure where the surface is continuously or almost continuously exposed to water-saturated air accompanied by very frequent or continuous condensation.

condensation polymer—polymer resulting from linking of monomers where water or other simple molecules are split off.

condensation resins—any of the alkyd, urea-formaldehyde, or phenol-aldehyde resins.

condensers—device used to cool a vapor to cause it to condense to a liquid. Also applies to shell-and-tube heat exchangers with steam condensing in the tubes and utilizing cooling water on the shell.

conductance—reciprocal of electrical resistance.

conduction—transmission of heat, light, sound, or electric charge.

conduction, electrical—passage of electric charge through a substance under the influence of an electric field.

conductivity—ability of a substance to conduct heat or electricity.

conductometric titration—titration in which the electrical conductivity of the reaction mixture is continuously monitored as a reactant is added.

conductor, electrical—substance or medium that conducts an electric charge.

conductor, thermal—substance that has a relatively high thermal conductivity because of its high concentration of "free electrons." *See also* **thermal conductivity**.

confined aquifer—aquifer bounded both above and below by confining strata or different formation layers.

congeal—to change from a liquid or semisolid state to a solid or rigid state.

congeners—elements belonging to the same group in the periodic table. Also a member of the same kind, class, or group. Also the by-products of reactions.

conjoint action—metal corrosion failure mechanism in which chemical and physical stresses combine to give more rapid failure than would be expected by their simple additive effects, i.e., a synergistic effect on failure. Types of corrosion attack that are often exacerbated by physical stresses to produce a conjoint action failure include bimetallic or galvanic corrosion, intergranular or intercrystalline corrosion, exfoliation corrosion, crevice corrosion, weld decay, stress corrosion, corrosion fatigue and fretting, filiform corrosion, hydrogen embrittlement, liquid-metal corrosion, and selective corrosion.

conjugated—double or triple bonds in a molecule separated by one single bond, e.g., $H_2H=CH–CH=CH_2$.

connate water—fossil water produced with oil. Also water laid down and entrapped with sedimentary deposits. Distinguished from migratory water, which is water that has flowed into deposits after they were laid down.

consensus standard—any standard developed according to a consensus agreement or general opinion among representatives of various interested or affected organizations and individuals.

consolute liquids—liquids that are miscible in all proportions under specified conditions.

consolute temperature—temperature at which two partially miscible liquids become fully miscible as the temperature is increased.

constant boiling mixture—*see* **azeotropic system**.

constantan—alloy of 50 to 60% copper and 40 to 50% nickel frequently used as one of the metals in thermocouples because its electrical resistance varies only very slightly with temperature (around room temperature).

constitutional diagram—graphical representation of the compositions, temperatures, pressures, or combinations of these at which the heterogeneous equilibria of an alloy system occur. Also called phase diagram.

constitutional formula—illustration of the composition of a chemical compound by displaying the individual atoms and radicals joined together by valency linkages.

constitutive property—property that depends on the constitution or structure of a molecule.

contact angle—angle formed between a solid surface and the tangent to a liquid drop on that surface at the line of contact between the liquid, the solid, and the surrounding phase (usually vapor or air), and measured through the liquid.

contact corrosion—term, primarily used in Europe, to describe galvanic corrosion. Also refers to the acceleration of metal corrosion from contact by nonmetals (e.g., wood, paper, polymers, concrete) in wet or humid environments due to the migration of corrosive ions from the nonmetal to the surface of the metal, or the nonmetal providing a source of water retained by either physical absorption or by deliquescent compounds within the nonmetal.

contact inhibitors—corrosion inhibitors that are impregnated in wrapping paper to protect manufactured metal components during temporary storage or in transit. Examples include sodium benzoate and sodium nitrite for steel, benzotriazole for copper and brass, and sodium chromate and sodium metasilicate for aluminum.

contact plating—metal plating process wherein the plating current is provided by galvanic action between the work metal and a second metal without the use of an external source of current.

contact potential—potential difference at the junction of two dissimilar metals or alloys.

contaminant—any undesirable or unacceptable physical, chemical, biological, or radiological substance or matter in or on a product or material.

contamination—presence of a substance not intentionally incorporated in or on a product or material.

continuous injection—method of continuously feeding a corrosion inhibitor in once through systems where batch treatment cannot be effectively or economically practiced. Used for once-through water supplies, oil field injection water, once-through cooling water, etc.

continuous phase—medium or continuum in which a dispersed phase is contained.

controlled cavitation water-jetting—water-jetting process for cleaning based on the principle of cavitation.

convection—process by which heat is transferred from one part of a fluid to another by movement of the fluid itself.

conventional current—convention that treats any electrical current as a flow of positive charge from a region of positive potential to one of negative potential. The real motion however, in the case of electrons flowing through a metal conductor, is in the opposite direction, from negative to positive.

conventional mud—drilling fluid containing essentially clay and water.

conversion coating—coating formed on metal either naturally by reaction with its environment or which is man-made using chemical or electrochemical treatment. *Also called* **chemical-reaction coating**. Examples include: anodizing of aluminum, magnesium, and titanium alloys; chromate coatings on zinc, aluminum, and cadmium; phosphate coatings on iron and steel, aluminum, and zinc; oxide bluing of iron and steel; oxide coatings on cadmium, copper, iron, steel, and zinc alloys; and pack cementation to form diffusion coatings on various metals. *See also* **blackening** and **electrochemical conversion coating**.

conversion factor—exact relationship between two quantities expressed as a fraction. Used to convert one set of units to another.

convertible coating—coating undergoing an irreversible transformation after its film formation to a film insoluble in the solvent from which it was deposited.

coolant—fluid used to remove heat. Liquid that circulates through a cooling system of heat-generating equipment and transfers heat from the equipment through some type of heat exchanger to the air.

cooling pond—reservoir used to cool a recirculating cooling water by natural evaporation, radiation, and convection for later reuse of the water in cooling, for some other purpose, or for disposal.

cooling range—cooling tower term describing the difference, in degrees Fahrenheit or Celsius, between the hot water returning to the tower and the cold water leaving it.

cooling towers—open structures wherein cooling waters are recirculated to give up their heat to the atmosphere and thus be reused for cooling. As part of the process, water evaporation is significant. Cooling towers are classified by the method used to induce air flow (either natural or mechanical draft) and by the direction of the air flow (either counterflow or crossflow relative to the downward flow of water).

cooling water deposits—deposits in cooling water systems that are classified as either foulants or scales. Fouling deposits are usually from suspended material and are typically composed of metal oxides, microbiological slime, oil, mud, siliceous silt, process contaminants, and chemical treatment residues. Scale deposits are the crystalline and noncrystalline deposits formed from dissolved materials that have precipitated because their respective solubilities have been exceeded. They are usually composed of compounds such as calcium carbonate or sulfate, and various silicate salts.

cooling waters—next to its agricultural use, the biggest use of available fresh water is in industry where it is mainly used for cooling. Cooling waters can be classified as: (1) once-through systems where the water passes through the cooling system only once, picks up heat, and returns to a receiving body of water; (2) closed recirculating systems where the water is circulated in a closed loop with negligible evaporation or exposure to the atmosphere to pick up heat and pass it to a heat exchanger before returning for more heat pickup; and (3) open recirculating systems, which utilize a cooling tower or evaporation pond to dissipate the heat it removes from a process or product. Because this latter system is open to the atmosphere, evaporation loss is significant and fresh makeup water must be incorporated to balance the loss.

coordinate covalent bond—single covalent bond in which one atom contributes two electrons to the formation of the bond.

coordinated phosphate—boiler treatment using phosphate buffers to avoid hydroxyl alkalinity.

coordination compound—compound with a central atom or ion bound to a group of ions or molecules surrounding it. They are usually derived by addition from simpler inorganic substances. *Also called* **complex compound**.

coordination number—number of groups, molecules, atoms, or ions surrounding a given atom or ion in a complex or crystal.

copolyesters—randomized block copolymers composed of polyester crystalline hard segments and amorphous glycol soft segments.

copolymer—polymer formed from two or more types of monomers. *See also* **graft polymer**.

copper-accelerated salt spray (CASS) test accelerated corrosion test used for evaluating electrodeposits and anodic coatings on aluminum.

copper displacement test—test for cleanliness of steel surfaces. The metal is immersed in an aqueous solution of copper sulfate, where a displacement reaction will take place depositing a uniform copper coating that forms quickly if the surface is clean of grease or other contaminants.

copper salts—compounds that find use as algicides and bactericides in water systems. Cupric sulfate ($CuSO_4 \cdot 5H_2O$), is the major example. Use of these salts, however, may increase corrosion of any metals in the system more active than copper.

copper staining—product of copper corrosion that washes down and stains another surface. Also the discoloration of copper due to its corrosion.

copper strip corrosion test—evaluation of the tendency of a product to corrode cupreous metals. See ASTM D 130 for details.

copper sulfate test—test for presence of mill scale on an iron or steel surface. Surface is swabbed with a 5 to 10% copper sulfate solution. Copper color formation indicates absence of mill scale.

corona—cracks on the surface of a material caused by ozone.

Corrator—trademark for an instrument that measures instantaneous corrosion rates by utilizing an electrochemical phenomenon known as linear polarization. It involves measuring the current required to change the electrical potential of a metal specimen corroding in an electrolyte by a few millivolts.

corrodkote test—accelerated corrosion test for electrodeposits. See ASTM B 380 for details.

corrosion—chemical or electrochemical reaction between a material, usually a metal, and its environment that produces a deterioration of the material and its properties. The term corrosion is also sometimes used in connection with the destruction of body tissues by strong acids and bases.

corrosion allowance—design dimensioning of a material so that its thickness is increased by an amount that will compensate for expected lifetime corrosion loss.

corrosion control processes—classified by the patterns of their anodic and cathodic polarization curves as being one of four types: anodic, cathodic, mixed anodic-cathodic, or resistance-controlled.

corrosion coupons—most widely used tools to monitor corrosion in process equipment and laboratory test procedures. The coupons can be made in any size, shape, or configuration from virtually any metal or alloy to allow for, after retrieval, visual examination, physical measurement, and chemical analysis and to study different types of corrosion mechanisms.

corrosion current—corrosion rate; usually expressed in A/cm^2.

corrosion damage—the corrosion effect that is considered detrimental to the function of the metal, the environment, or the technical system of which these form a part.

corrosion effect—change in any part of the system caused by corrosion.

corrosion embrittlement—severe loss of ductility of a metal resulting from corrosive attack, usually intergranular corrosion, and often not visually apparent.

corrosion engineering—application of science and art to prevent or control corrosion damage economically and safely.

corrosion-erosion—*see* **erosion-corrosion**.

corrosion fatigue—conjoint action of corrosion and fatigue (cyclic stressing) to cause metal fracture.

corrosion fatigue limit—maximum stress endured by a metal without failure in a stated number of stress applications under defined conditions of corrosion and stressing. Also called corrosion fatigue strength.

corrosion index—measurement of the corrosivity of water. *See also* **Langelier saturation index** and **Ryzner stability index**.

corrosion-induced deformation—strains in metallic wall panels caused by expansions resulting from corrosion.

corrosion inhibitor—chemical substance, or combination of substances, that when added in the proper concentration and form to an environment reduces the corrosion rate of a material exposed to that environment.

corrosion inhibitor classifications—any of several inhibitor classifications, including by their composition, mechanism of action, or form. A common classification is: (a) anodic inhibitors that function by forming an insoluble protective film at anodic sites on the surface or by adsorbing at anodic sites on the metal; (b) cathodic inhibitors that function by forming an insoluble protective film at cathodic sites on the surface or by adsorbing at these sites; (c) organic inhibitors that adsorb over the entire surface thus blocking both anodic and cathodic reactions; (d) vapor phase inhibitors that are used mainly in closed systems and function by volatilizing to carry them throughout and to the surface where they are adsorbed; (e) oxygen removers, which are used mainly in closed aqueous systems, and that function by reacting with dissolved oxygen in water to reduce or eliminate the oxygen as a corrodent.

corrosion inhibitor types—four major types of corrosion inhibitors based on the mechanism of their operation include the barrier layer formers (oxidizers, adsorbed layer formers, and conversion layer formers), neutralizers, scavengers, and miscellaneous (scale inhibitors and biological growth inhibitors).

corrosion potential—rest potential of a corroding metal in an electrolyte relative to a reference electrode measured under open-circuit conditions. It is electrochemically established as the potential at the intersection of the anodic-cathodic polarization curves. *Also called* rest potential, **open-circuit potential**, and freely corroding potential.

corrosion products—products resulting from the corrosion of a material.

corrosion rate—speed with which corrosion progresses.

corrosion resistance—ability to withstand corrosion in a given corrosive environment.

corrosion scientist—scientist concerned with the study of corrosion mechanisms through which a better understanding is obtained of the causes of corrosion and means for preventing or reducing resulting damage.

corrosion system—system of one or more metals and all parts of the environment that influence corrosion.

corrosion tendency—reactivity of a metal with its environment as measured by the Gibbs free-energy change. The more negative the value the greater the tendency for corrosion.

corrosive wear—wear in which chemical or electrochemical reaction with the environment is a significant contributor.

corrosivity—tendency of an environment to cause corrosion.

Corrosometer—trademark for an instrument for monitoring corrosion by measuring the increase in electrical resistance of a metal specimen probe as its cross-sectional area is reduced by corrosion. The probe is made of an alloy identical or similar to the metal in the system.

Cor-Ten steel—trademark for atmospherically slow-rusting, copper-rich low-alloy steel nominally containing 0.41% copper, 0.51% nickel, 1.0% chromium, and small amounts of other elements. *See also* **weathering steels**.

corundum—natural aluminum oxide (Al_2O_3), used as an abrasive.

cosmic rays—highly penetrating radiations that strike the earth and originate in interstellar space. Classed as primary, coming from the assumed source, and secondary, being induced in upper atmospheric nuclei by collisions with primary cosmic rays.

cosolvent—solvent used to improve the mutual solubility of other ingredients or to cause two immiscible liquids to mix. Also called coupling agent.

coulomb—quantity of electricity that is transmitted through an electric circuit in one second when the current in the circuit is one ampere.

counter current—process in which a material is caused to flow against another stream.

counter electrode—*see* **auxiliary electrode**.

couple—cell developed in an electrolyte resulting from electrical contact between two dissimilar metals. *See also* **galvanic corrosion**.

coupling—chemical reaction in which two molecules join together.

covalent bond—chemical bond that results when electrons are shared between two nuclei; the overlap of atomic orbitals from two atoms.

coverage—ambiguous coating term generally used to refer to hiding power or spreading rate. *See also* **hiding power** and **spreading rate**.

covering power—ability of a plating solution, under specified plating conditions, to deposit metal on the surfaces of recesses or deep holes.

CP—*abbreviation for* **cathodic protection**.

CPVC—*abbreviation for* **critical pigment volume concentration**.

cracking—fracture of a metal in a brittle manner. Also the process of breaking down chemical compounds by heat (pyrolysis). Also an oil-refining process that breaks large molecules into smaller ones. Also any fissure in a coating other than a mechanical cut.

cratering—small round depressions in a paint film that may or may not expose the underlying surface.

crawling—defect in which a wet coating film recedes from a small area to form an uneven surface after application.

crazing—development of fine cracks on the surface of a material.

creaming—settling or rising of the particles of the dispersed phase of an emulsion as observed by a difference in color shading of the layers formed.

creep—continuous deformation of a solid material, usually a metal, under a constant stress that is well below its yield point.

creep strength—stress that causes a given elongation of a material.

creosol—2-methoxy-4-methylphenol ($C_8H_{10}O_2$). It is one of the active constituents of creosote. Not to be confused with cresol.

creosote (coal tar)—translucent brown to black oily liquid mixture of phenols obtained by distilling tar recovered from the destructive distillation of wood. Used as a wood preservative, insecticide, and disinfectant.

cresol—one of a possible three isomers of hydroxymethyl benzene (C_7H_8O) obtained from coal tar. *Also called* **cresylic acid**. A colorless, yellowish, brownish, or pinkish liquid of characteristic phenolic odor. Used as a disinfectant.

cresylic acid—commercial mixture of phenolic materials that may include phenol, various cresols, xylenols, and other alkylated phenols. *See also* **cresol**.

crevice—narrow crack or opening.

crevice corrosion—localized corrosion resulting from the formation of a concentration cell in a crevice.

crevice types—crevices are categorized as either man-made or naturally occurring. Man-made crevices may result from a deliberate design feature such as from the use of a fastener or gasket; may be formed during fabrication or assembly; or may result from the use of sealants, coatings, or even grease. Naturally occurring crevices may be formed by the presence of debris, sand, barnacles, etc., on a surface.

C-ring—C-shaped specimen used in quantitatively determining the susceptibility to stress-corrosion cracking of all types of alloys in a wide variety of product forms. See ASTM G 38 for details.

critical anodic current density—maximum anodic current density observed in the active region for a metal or alloy electrode that exhibits active-passive behavior in an environment.

critical current density—current density above which a new reaction occurs.

critical humidity—relative humidity above which the atmospheric corrosion rate of some metals increase sharply.

critical pigment volume concentration (CPVC)—that level of pigmentation in the dry paint film where just sufficient binder is present to fill the voids between the pigment particles.

critical pitting potential—lowest value of oxidizing potential at which pits nucleate and grow.

critical point—point where two phases, which are continually approximating each other, become identical and form only one phase.

critical pressure—pressure of a fluid when it is in its critical state, i.e., at critical temperature and critical volume. Above this pressure, the fluid no longer has the properties of a liquid regardless of further increase in pressure.

critical state—state of a fluid in which the liquid and gas phases both have the same density.

critical temperature—temperature above which a gas cannot be liquefied by pressure alone. The pressure under which a substance may exist as a gas in equilibrium with the liquid at the critical temperature is known as the critical pressure.

critical-toxicity range—interval between the highest concentration at which all test organisms survive and the lowest concentration at which all test organisms die within the test time period.

critical volume—volume of a fixed mass of a fluid in its critical state.

crizzle—surface imperfection on a material in the form of a multitude of fine fractures.

crocus cloth—iron oxide-coated abrasive cloth used for polishing.

cross-hatch test—test to demonstrate adhesion characteristics of a paint- or powder-coated surface. It is performed by scribing a cross-hatch pattern through the coating, covering the scribed area with tape, then sharply removing the tape and viewing the degree of remaining adhered paint.

cross link—chemical bond bridging one polymer chain to another.

cross linkage—short side chain of atoms linking two longer chains in a polymeric material.

crosslinking—setting up of chemical links between the molecular chains of polymer molecules to form a three-dimensional or network polymer, generally by covalent bonding.

crude oil—naturally occurring hydrocarbon mixture, generally in a liquid state, which may also include compounds of sulfur, nitrogen, oxygen, metals, and other elements. *Also called* **petroleum**.

cryogenic coating removal—use of liquid nitrogen to remove an organic coating by embrittling the coating then tumbling or blasting with plastic pellets.

cryohydrate—solid that separates when a saturated solution freezes. It contains the solvent and solute in the same proportions as they were in the saturated solution.

crystal—macroscopic solid composed of atoms, ions, or molecules arranged in a pattern that is periodic in three dimensions, i.e., exhibits some degree of geometrical regularity or symmetry.

crystal distortion—any change in the normal crystalline structure of a crystal or of a precipitated solid that is crystalline.

crystalline solid—solid whose atoms, ions, or molecules are arranged in an orderly, regular, three-dimensional pattern.

crystal modifiers—chemicals that are used as deposit control agents acting to modify the crystal structure of scale; widely used in industrial water systems, especially cooling water and boiler water. They distort scale crystal growth so that precipitates are structurally weak, more readily dispersed, and less prone to adherent buildup. Lignin, tannin, and various synthetic polymers are frequently used as crystal modifiers.

C-stage—final stage in the reaction of certain thermosetting materials whereupon they have become practically insoluble and infusible. *See also* **A-stage** and **B-stage**.

cup grease—generally a lime-base grease suitable for use in grease cups.

cupreous—concerning, resembling, or containing copper. Also used to describe copper or copper alloy materials.

cupric—copper in its higher (+2) oxidation state.

cupronickel—corrosion-resistant alloy of copper and nickel containing up to 45% nickel.

cuprosolvency—electrochemical general corrosion of copper in domestic water systems characterized by staining of plumbing fixtures and causing water to color a blue/green. Cuprosolvency occurs most frequently in soft, low alkalinity, low mineralization water of pH 7 or lower.

cuprous—copper in its lower (+1) oxidation state.

cure—process to change the physical properties of a material by chemical reaction or by the action of heat or a catalyst. Also the process by which paint is converted from a liquid to a solid state, usually by means of condensation, polymerization, or vulcanization using heat or catalysts. Also the process used to change the properties of a polymeric system into a final, more stable, and usable condition by the use of heat, radiation, or reaction with chemical additives.

curing temperature—temperature at which a material is subjected to curing.

curling—coating defect similar to crawling. *See also* **crawling**.

current—net transfer of electric charge per unit time. Also called electric current. The instantaneous rate at which a positive charge passes through a surface.

current density—quantitative current flowing through an electrolyte per unit area of an electrode surface. Also the current flowing through a conductor per unit cross-sectional area. Usually expressed in amperes per square meter.

current efficiency (Faradaic efficiency)—proportion, expressed as percentage, of the current, that is effective in carrying out a specified process in accordance with Faraday's law. It can be determined for a given metal M by the equation: current efficiency$_M$ is equal to the quantity of electricity producing M divided by the total quantity of electricity used.

curtain coating—method of applying paint by moving the object to be painted through a falling curtain of paint.

cut-back asphalt—asphalt that has been blended with petroleum distillates.

cutting fluid—any fluid (liquid, gas, or mist) applied to the working part of a tool in a cutting operation to promote more efficient machining by improving cooling, lubrication, and, possibly, corrosion protection.

cutting oil—cutting fluid composed of oil of petroleum, animal, or vegetable origin, either singly or in combinations. Soluble chlorine, sulfur, or phosphorus additive agents can be incorporated to improve lubrication.

cyaniding—introducing carbon and nitrogen into a solid ferrous alloy by contact with molten cyanide compositions.

cycles of concentration—term used in an evaporating water system. It is the ratio of the concentration of dissolved substances in the evaporated water to their concentration in the original or makeup water before evaporation. Cycles of concentration are usually measured in the blowdown. *See also* **blowdown** and **wastage**.

cyclic corrosion testing (CCT)—an accelerated laboratory testing method performed in a closed chamber where wet and dry environmental conditions simulating actual atmospheric exposure are artificially simulated. Also the abbreviation for comfort cooling tower.

cycloalkanes—cyclic saturated hydrocarbons containing a ring of carbon atoms joined by single bonds and having the general formula C_nH_{2n}, e.g., cyclohexane (C_6H_{12}).

cyclone—mechanical dust collection device that operates by spinning the collected material within the device using centrifugal force to direct the dust to the outside wall of the separator. Gravity and mechanical internal deflectors direct the dust-laden air in a downward spiral and discharge the dust out the cyclone cone bottom. The internal airstream turns and spirals up the center of the cyclone and out the top.

D

Dalton's law—total pressure of a mixture of gases or vapors is equal to the sum of the partial pressures of its components.

damp—wet; not dry.

dangerous corrosion inhibitor—corrosion inhibitor that, when used at below a critical concentration, accelerates rather than retards corrosion.

Daniell cell—galvanic device consisting of a zinc anode immersed in a zinc sulfate ($ZnSO_4$) solution and a copper cathode immersed in a solution of copper sulfate ($CuSO_4$) with the two solutions separated by a diaphragm. The emf of the cell is 1.10 volts.

data—information collected when an experiment is conducted.

DC (d.c.)—*abbreviation for* **direct current**.

deactivation—process of removing active constituents from a corroding medium, usually oxygen, by the controlled corrosion of an expendable metal or by other chemical means to make the medium less corrosive. The term also refers to the partial or complete reduction in the reactivity of a substance as in the poisoning of a catalyst.

deactivation of water—pretreatment of water to remove its dissolved oxygen by chemically reacting the oxygen, e.g., by flowing the heated water over a large surface of steel to corrode it thereby consuming the oxygen, by reaction with sodium sulfite or hydrazine.

deadhesion—process whereby organic coatings may separate from metallic substrates. It includes the loss of adhesion from paint application to a wet substrate, osmotic and cathodic blistering, cathodic delamination, thermal cycling, and anodic undermining.

deaeration—process of removing dissolved gas from a liquid environment by a deactivating or heating process. Usually refers to removal of oxygen and carbon dioxide from boiler feedwater. In water-treatment systems, deaeration is usually accomplished by spraying steam-heated water or flowing the boiler feedwater over a large surface countercurrent to steam. Dissolved gases may also be removed by lowering the pressure using a mechanical pump thus effecting vacuum deaeration.

deagglomeration—process of breaking down masses of particles that are held together by relatively weak cohesive forces. Usually refers to a physical process.

dealkalization—any process for reducing the alkalinity of water.

dealloying corrosion—selective removal, by corrosion, of a metallic constituent from an alloy. *Also known as* **parting corrosion** and **selective leaching**.

dealuminification—form of leaching of aluminum from aluminum bronzes and usually occurring in seawater environments.

debris—particles detached in a wear or erosion process.

debye—unit of electric dipole moment in the electrostatic system and used to express dipole moments of molecules.

Debye-Huckel theory—explains the nonideal behavior of electrolytes. It assumes that electrolytes in solution are fully dissociated and that nonideal behavior arises because of electrostatic interactions between the ions.

decantation—pouring or siphoning off the supernatant liquid above a precipitate or sediment or from two immiscible liquids.

decarburization—loss of carbon from the surface layer of a carbon-containing alloy due to its reaction with the environment.

decay—decomposition of wood caused by wood-destroying fungi.

dechlorination—removal of dissolved chlorine from water. Commonly accomplished with activated carbon or by chemical-reducing agents such as sulfur dioxide, sodium sulfite, sodium bisulfite or sodium thiosulfate.

deck paint—paint having a high degree of resistance to mechanical wear and usually used on porch floors and ship decks.

decobaltification—corrosion and selective leaching of cobalt from cobalt-base alloys or from cemented carbides.

decomposition—chemical reaction in which a compound breaks down into simpler compounds or into its elements.

decomposition potential (or voltage)—potential of a metal surface necessary to decompose the electrolyte of a cell or a substance in it.

decomposition voltage—*see* **decomposition potential**.

decorative painting—painting for cosmetic appearance.

deep groundbed—one or more anodes placed vertically at 50 feet or more below the surface of the earth in a drilled hole to supply cathodic protection for underground or submerged metallic structures.

defect—discontinuity in a crystal lattice.

deflocculation—state or condition of dispersion of a solid in a liquid in which each solid particle remains independent and unassociated with adjacent particles.

defoamer—additive used to reduce or eliminate foam.

deformation—any change in form or shape of a body produced by a stress or force.

degassing—removal of dissolved or absorbed gases from a liquid or solid.

degradation—deleterious change in the physical or chemical properties of a material.

degreasing—removal of grease or oil from a surface using a chemical compound or solution.

degree—division on a temperature scale.

degrees of freedom—number of independent parameters required to specify the configuration of a system (usually pressure, temperature, and concentrations of the components). Also the least number of independent variables required to define the state of a system in the Gibbs phase rule.

dehumidify—to reduce the quantity of water vapor from air and other gases within a given space. Usually accomplished by cooling below the dew point so that part of the water vapor is condensed and/or by adsorption of moisture by various chemical desiccants.

dehydration—removal of water from a substance, system, or chemical compound. Also a chemical reaction in which a compound loses hydrogen and oxygen in the ratio 2:1, respectively.

dehydrogenation—reaction that results in the removal of hydrogen from an organic compound.

deicing salts—usually refers to sodium or calcium chloride, sometimes mixed with fine sand to act as an abrasive, which is spread on a road to melt snow and ice as a safety measure.

deionization—process of removing ions from water, most commonly by an ion-exchange process where cations and anions are removed independently of one another.

delamination—separation of layers in a multilayered structure.

delignification—dissolution of the lignin portion of cooling tower wood, usually by alkaline agents, oxidizing agents, and/or microorganisms, resulting in a fibrous, weakened structure.

deliquescence—absorption of water from the atmosphere by a hygroscopic solid to such an extent that a concentrated solution of the solid eventually forms.

demineralization—removal of dissolved mineral matter from water. *Also called* **desalination**.

demulsibility—ability of an oil to separate readily from water or to resist its emulsification.

dendrite—crystal, usually formed during solidification or sublimation, which is characterized by a treelike pattern composed of many branches.

denickelification—selective leaching of nickel from nickel-containing alloys. *See also* **dealloying corrosion**.

density—mass of a unit volume of material.

denting—physical deformation of a metal or alloy due to the pressure (stress) of corrosion products developed within it or from adjacent contacting metal or alloy. *Also called* **wedging action**.

deoxidizing—(1) removal of oxygen from molten metal using suitable deoxidizers. (2) removal of oxide film from a metal surface by chemical or electrochemical reaction.

depletion—selective removal of one component of an alloy from the surface or from grain-boundary regions.

depolarization—elimination or reduction of polarization by physical or chemical means resulting in increased corrosion of a metal.

depolarizer—substance that produces depolarization.

deposit—foreign substance coming from the environment and adhering to the surface of a material.

deposit attack—*see* **deposit corrosion**.

deposit control agents—chemicals added to cooling waters to reduce their deposit buildup. Typical deposit control agents include chelants such as EDTA and NTA, lignosulfonates, polyphosphates, polyacrylates, polymethacrylates, maleic anhydride copolymers, polymaleic anhydride, phosphate esters, and phosphonates.

deposit corrosion (deposit attack)—corrosion occurring under or around a discontinuous deposit on a metallic surface. *Also called* **poultice corrosion**. It is a form of concentration cell corrosion.

deposition corrosion—form of galvanic corrosion in which corrosion of metal upstream can produce ions carried downstream to deposit on a more active metal thus causing dissimilar metals or galvanic corrosion leading to pitting where it is deposited.

derived units—SI units that are obtained from combinations of the seven base units. *See also* **SI**.

dermatitis—inflammation of the skin.

desalination—removal of inorganic dissolved solids, usually sodium chloride, from water. *Also called* **demineralization**.

desalting—removal of impurity salts from water, usually by electrodialysis, reverse osmosis, or distillation.

descaling—removing caked rust or the thick layer of oxides formed on some metals when heated to elevated temperatures using chemical or mechanical means or a flame as in flame cleaning.

desiccant—substance used to absorb water vapor from the surrounding air by either chemical or physical means to maintain a low relative humidity in a container or space. Physical desiccants are the most widely used and include silica gel, activated alumina, bentonite clays, and synthetic zeolites. Common chemical desiccants include lime, CaO, and a mixture of CaO and Na_2O (soda lime).

desiccator—container used in drying enclosed substances or for keeping them free from moisture.

desorption—removal of adsorbed atoms, molecules, or ions from the surface of a material or substance.

destructive distillation—decomposition of organic compounds by heating in the absence of air.

desulfovibrio microorganisms—various species of sulfate-reducing bacteria involved in the anaerobic corrosion of iron and steel.

detergent—surface-active agent possessing the ability to clean soiled surfaces. Soap is the original example. Various synthetic detergents, which are soapless, are made from petrochemicals, e.g., dodecylbenzenesulfonate. Detergents are classified either as anionic, nonionic, or cationic depending on whether their active part is a negative, neutral, or positive ion, respectively.

deterioration—permanent impairment of the chemical or physical properties of a material.

deuterium—isotope of hydrogen with a mass number 2.

deuterium oxide—*see* **heavy water**.

deuteron—nucleus of the deuterium atom or the ion of deuterium. Its structure contains one neutron and one proton.

deviation—variation from a specified value. Also term used to define allowable upper and lower limits from a specified value.

devitrification—crystallization of an amorphous substance.

dew—*see* **precipitation**.

dewater—any process to separate water from sludge to produce a cake that can be handled as a solid.

dew point—temperature to which water vapor must be reduced to obtain saturation vapor pressure, i.e., 100% relative humidity. Also temperature at which moisture condenses.

dew-point corrosion—corrosive attack in the low-temperature section of fossil-fuel power plant combustion equipment resulting from acidic flue gas vapors that condense and cause corrosion damage.

dew-point hygrometer—*see* **hygrometer**.

dezincification—corrosion in which zinc is selectively leached from zinc-containing alloys; most commonly seen with copper-zinc alloys that have dwelled for extended time in water containing dissolved oxygen. Dezincification takes place either in localized areas on the alloy surface (*called* **plug-type dezincification**), or uniformly over the surface (called layer-type dezincification).

DFI—abbreviation for driving force index. *See also* **McCauley's driving force index**.

DFT—*abbreviation for* **dry film thickness**.

dialysis—process of separating solutes from a solution and which depends on differences in their diffusion rates across a semipermeable membrane.

diaphragm—porous or permeable membrane separating anode and cathode compartments of an electrolytic cell from each other or from an intermediate compartment.

diatomaceous earth—infusorial earth composed of siliceous skeletons, about 88% silica (SiO_2), of diatoms and being very porous. Used in filtration and clarifying.

diatoms—particular group of unicellular algae predominating in fresh and salt water environments as well as in soils. They have cell walls composed of pectin impregnated with silica.

dicarboxylic acids—group of compounds containing two carboxylic acid groups (–COOH).

dielectric—nonconductor of electric charge in which an applied electric field causes a displacement of charge but not a flow of charge.

dielectric coatings—pipeline coatings that control corrosion by isolating the external metallic surface of underground or submerged piping from the environment. They reduce cathodic protection requirements and improve protective current distribution.

dielectric constant—specific property of an insulating material that is defined and measured by the ratio of electrical capacity of a condenser having that material as the dielectric to the capacity of the same condenser having air as the dielectric. The dielectric constant of air is unity. Also known as the relative permittivity.

dielectric loss—time rate at which electric energy is transformed into heat in a dielectric when it is subjected to a changing electric field.

dielectric shield—term used in a cathodic protection system. It is an electrically nonconductive material, such as a paint or varnish coating or plastic sheet or pipe that is placed between an anode and an adjacent cathode to avoid current wastage and to improve current distribution, usually on the cathode.

dielectric strength—voltage gradient at which dielectric failure of an insulating material occurs. *Also called* **electric strength**.

dielectric value—*see* **dielectric strength**.

diesel fuel oil—any petroleum liquid suitable for the generation of power by combustion in compression ignition (diesel) engines.

differential aeration cell—corrosion cell resulting from the potential difference caused by differences in oxygen concentration along the surface of a metal in an electrolyte. It is one example of a concentration cell. *Also known as* **oxygen concentration cell**. *See also* **concentration cell**.

differential temperature cell—corrosion cell whose components are electrodes of the same metal and immersed in an electrolyte of the same initial composition, but which electrodes are at a different temperature.

diffraction—phenomena produced by the spreading of waves around and past obstacles that are comparable in size to their wavelength.

diffusion—displacement of atoms, molecules, or ions in a liquid or gas under the effect of a concentration difference. It is the process by which different substances mix as a result of the random motions of their component atoms, molecules, and ions.

diffusion coating—alloy coating process produced by applying heat to one or more metal coatings deposited on a basis metal. Also process by which a basis metal is exposed to a gaseous or liquid medium containing another metal or alloy to bring about diffusion of the latter into the basis metal to change its surface composition. Process is sometimes termed surface alloying.

diffusion limited current density—current density that corresponds to the maximum transfer rate that a particular species can sustain due to the limitation of diffusion.

digester—jacketed reaction vessel used to soften, decompose, or cook substances at high temperature and pressure.

dilation—an expansion, stretching, or increase in volume.

diluent—gas, liquid, or solid used to reduce the concentration of an active ingredient in a system.

dilute—solution having a relatively low concentration of solute.

dilution—stated volume of solvent in which a given amount of solute is dissolved.

dimer—condensation product of two identical molecules (monomers).

dimer acids—liquid polycarboxylic acids produced by polymerization of unsaturated fatty acids. It is a C_{36} dicarboxylic acid that may contain minor amounts of C_{18} and C_{54} monocarboxylic and tricarboxylic acids, respectively, as impurities.

DIN—abbreviation for Deutsche Industrie-Norm (German Industry Standard).

dip coating—coating process in which a substrate is immersed in a solution or dispersion containing the coating material and then withdrawn.

diphase cleaner—composition that produces two phases in a cleaning tank, a solvent layer and an aqueous layer, which cleans by both solvent action and emulsification.

dipolar ion—chemical species that contains a positive charge and a negative charge; a zwitterion.

dipole—pair of electric charges or magnetic poles of equal magnitude but opposite sign or polarity that are separated by a small distance.

direct current (DC or d.c.)—electric current in which the net flow of charge is in one direction only. *Compare with* **alternating current**.

dirt—small particle of foreign material embedded in or on a material. *See also* **soil**.

disbondment—destruction of adhesion between a coating and the surface coated.

discharge—(1) release of electric charge from a capacitor in an external circuit. (2) conversion of the chemical energy stored in a secondary cell into electrical energy.

discontinuity—any interruption in the normal physical structure or configuration of a part, such as cracks, laps, seams, inclusions, or porosity.

discontinuous phase—in an emulsion, it is that phase which is broken down into droplets or particles and dispersed throughout the continous phase. *Contrast with* **continuous phase**. *See also* **internal phase**.

disinfection—use of energy or chemicals to destroy, neutralize, or inhibit the growth of pathogenic and other organisms. Does not necessarily mean destruction of all living organisms.

dislocation—linear imperfection in a crystalline array of atoms. It is the boundary between slipped areas of a crystal and unslipped regions.

dispersant—chemical that causes particles in a water system to remain in suspension.

dispersed phase—phase in an emulsion or suspension that is broken down into droplets or discrete particles and becomes dispersed throughout the other or continuous phase.

dispersing agent—substance that increases the stability of a suspension of particles in a liquid medium.

dispersion—suspension of one substance in another.

dispersion coating—paint in which binder molecules are present as colloidal particles. It is usually a high solids coating.

displacement corrosion—*see* **galvanic corrosion**.

displacement reaction—*see* **substitution reaction**.

disproportionation—type of chemical reaction in which the same compound is simultaneously reduced and oxidized.

dissimilar electrode cell—galvanic cell of the type illustrated by a dry cell, metal with surface impurities as a separate phase, cold-worked metal in contact with the same metal annealed, grain-boundary metal in contact with grains, etc., where anodic/cathodic reactions can take place when an electrolyte is present.

dissimilar soils—refers to soils of varying composition, e.g., coarseness and grain size, type (rock, loam, and clays), pH, and chemical constituents that can lead to buried pipeline galvanic corrosion.

dissociation—breakdown of a molecule or ion into smaller molecules or ions.

dissolved-gas-drive reservoirs—oil field reservoirs in which the oil is produced (recovered) by dissolved gas escaping and expanding within the oil forcing it to the surface for collection. *See also* **gas-cap-drive oil reservoirs, water-drive reservoirs,** and **oil-well pumps**.

dissolved gases—gases that enter into solution with a fluid and are neither free nor entrained gases.

dissolved matter—matter, exclusive of dissolved gases, that is dispersed in water to give a single homogeneous liquid phase.

dissolved oxygen—atmospheric oxygen in solution.

distensibility—ability to be stretched.

distillation—process of purifying or separating liquids by volatilizing and condensing them.

distilled water—water purified by distillation so as to be free from dissolved salts and other compounds. Distilled water in equilibrium with carbon dioxide from the air has a conductivity of about 0.8×10^{-9} siemens $\cdot$ cm^{-1}.

dithionate—*see* **suflinate**.

DOC—abbreviation for the Department of Commerce.

DOD—abbreviation for the Department of Defense.

DODISS—abbreviation for the Department of Defense Index of Specifications and Standards.

DOE—abbreviation for the Department of Energy.

dolomite—mineral having empirical composition of one mole of calcium carbonate and one mole of magnesium carbonate, i.e., $CaCO_3 \cdot MgCO_3$.

Donnan effect—rejection of diffusion of external ions by a semipermeable membrane because of high internal concentration of ions of the same charge.

DOT—abbreviation for the Department of Transportation.

double bonds—covalent bond in which four electrons are shared between two nuclei.

double decomposition (metathesis)—chemical reaction involving the exchange of radicals between reactants.

double-dip galvanizing—immersion of one-half of a structure at a time in a molten zinc bath. Used when the structure is too large to be immersed in one dipping.

double layer—interface between a metal and an electrolyte where an electrical charge separation is occurring.

double salt—crystalline salt in which there are two different anions and/or cations.

Dowtherm—trademark for eutectic mixture of phenyl ether and diphenyl that has a boiling point of 258 °C (497 °F). Used as a heat-transfer fluid.

drainage—conduction of electric current from an underground structure by using a metallic conductor.

drawing compound—die or blank lubricant used to increase die life and improve surface finish of metal product being drawn.

dressed lumber—lumber having one or more of its faces planed smooth.

drier—catalyst added to speed the cure or oxidation of oil-base paints, printing inks, or varnish. Common driers are the naphthenate salts of lead, manganese, and/or cobalt.

drift—entrained water in the stack discharge of a cooling tower. Also known as windage. Besides water, the drift contains any water minerals and treatment chemicals. In this regard, it differs from plume.

drift loss—*see* **windage loss**.

drilling—method of cutting a hole in solid metal using a rotary end-cutting tool with one or more flutes to promote chip removal and admission of cutting fluid. Also an abrasive cleaning method for tightly plugged heat-exchanger tubes where a hollow drill is used with a fluid connection to flush out freed material.

drilling mud (or fluid)—circulating fluid used in oil field rotary drilling to perform any or all of various functions required in the drilling operation. These include bringing cuttings from the bottom of the hole to the surface, controlling subsurface pressures, or cooling and lubricating the bit and drill string.

drilling mud (or fluid) corrosion—corrosion associated with using water-base drilling fluids for petroleum production. Its main environmental causes are associated oxygen, carbon dioxide, hydrogen sulfide, various ionic concentrations and low pH. Major forms of attack are varied and include crevice corrosion, deposit corrosion, corrosion fatigue, stress-corrosion cracking, erosion-corrosion, uniform corrosion, and galvanic corrosion.

drop—small body of liquid held together primarily by surface tension.

dropping point—temperature at which a grease passes from a semisolid to a liquid state under specified test conditions.

drossing—removal of refuse and impurities from a galvanizing bath.

dry-air storage—storage area whose air relative humidity is reduced to a level below the critical humidity so as to considerably reduce metallic corrosion rates.

dry-bulb temperature—temperature of air as indicated by a thermometer and corrected for radiation, if significant.

dry cell—primary or secondary cell in which the electrolytes are in the form of a paste.

dry corrosion—*see* **gaseous corrosion**.

dry etching—etching by means of sand blasting, grinding wheels, or wire brushes.

dry film thickness—depth of applied coating when dry.

dry hydrate—calcium hydroxide ($Ca(OH)_2$).

dry ice—solid carbon dioxide (CO_2); primarily used as a refrigerant.

drying oil—oil that, when exposed to air as a thin film, readily takes up oxygen from the air to change to a relatively hard, tough elastic substance.

dry rot—fungous disease of timber causing it to become brittle and crumble into powder. Occurs frequently as underspread damage in buildings. The causal organism is *Portia incrassata*.

dry wall—rigid interior wall board such as gypsum board or plywood.

dry/wet tower—cooling tower design for water conservation containing dry heat exchangers in conjunction with wet heat exchangers to optimize the cost of the system.

DSC—abbreviation for a thermal analytical technique known as differential scanning calorimetry.

DTA—abbreviation for a thermal analytical technique known as differential thermal analysis.

dual-phase steel—steel that has been heated to above 723 °C (1330 °F) into the ferrite and austenite region then quenched to produce a martensitic structure within a ferritic matrix resulting in ductility and which, on heating, the martensite will transform, producing carbides and a strong steel.

Dubpernell test—method for revealing cracks, pores, and other discontinuities in chromium deposits. The test involves electrodepositing copper on a chromium-plated surface from an acid-copper-sulfate solution. Copper is deposited only on those areas where cracks, pores, or other discontinuities exist.

ductile fracture—tearing of metal accompanied by appreciable gross plastic deformation and expenditure of considerable energy. *Compare with* **brittle fracture**.

ductile iron—cast iron that has been treated in the liquid state so as to cause substantially all its graphitic carbon to occur as spheroids or nodules in the as-cast condition. It exhibits ductility as a result.

ductility—ability of a material to deform plastically without fracturing.

dummy cathode—cathode in a plating solution that is not to be used after plating. Often used for removal or decomposition of impurities.

duplex stainless steel—chromium-molybdenum stainless steel composed of a mixture of austenite and ferrite.

duplicates—*see under* **statistical terms**.

Duralumin—trademark for a class of strong, lightweight aluminum alloys containing approximately 4% copper and usually lesser amounts of magnesium, manganese, and, sometimes, silicon.

Duriron—trademark for a high-silicon iron containing about 14.5% silicon and 0.95% carbon that is extremely corrosion resistant to many environments.

durometer—instrument for measuring hardness of rubber, plastics, or protective coatings.

dust—loose term applied to solid particles capable of temporary suspension in air or in other gases.

dust blasting—cleaning of the surface by the use of a very fine abrasive through a sandblast mechanism.

dye—substance used to impart color and that is generally an organic compound containing conjugated double bonds.

dynamic equilibrium—equilibrium that results when the rates of two opposing processes are equal.

dynamic friction—friction between two surfaces in relative motion. *Also called* sliding friction, **kinetic friction, and friction of motion.**

dynamometer—instrument used to measure a force, or a device used to measure the output power of an engine or motor.

dyne—absolute unit of force in the c.g.s. system that, acting upon a mass of 1 gram, will impart to it an acceleration of 1 centimeter per second squared.

E

early rusting—onset of a measles-like rusting appearance that occurs after a latex paint coating has dried to the touch. Usually brought on by a combination of (a) thin coating, (b) cool substrate temperature, and (c) high-moisture conditions.

earth pigments—colored natural pigments mined from the earth and processed to pigmentary utility. Also called natural or mineral pigments or colors.

earth's atmosphere—gas surrounding the earth. Composition of dry air at sea level is: nitrogen 78.08%, oxygen 20.95%, argon 0.93%, carbon dioxide 0.03%, neon 0.0018%, helium 0.0005%, krypton 0.0001%, and xenon 0.00001%. Air also contains water vapor and, in some localities, sulfur and nitrogen compounds, hydrogen peroxide, hydrocarbons, dust, and other matter.

ebullition—bubbling, effervescence, or boiling of a liquid.

EC50—estimated concentration that is expected to cause an effect in 50% of a group of organisms under specified conditions.

EC test—abbreviation for the electrolytic corrosion test.

ecology—study of the interrelationships between organisms and their natural environments, both living and nonliving.

economizer—heat exchanger in a furnace stack that transfers heat from the stack gas to the boiler feedwater.

ecosystem—biological community and the physical environment associated with it.

ED50—estimated dose that is expected to cause an effect in 50% of a group of organisms under specified conditions.

eddy current—current induced in a conductor situated in a changing magnetic field or moving in a fixed one. A current that runs contrary to the main current.

eddy current tester—device based on using eddy current to locate loss of metal or flaws in metal.

EDTA—abbreviation for ethylenediaminetetraacetic acid whose sodium salt is often used as a chelating agent.

effervescence—vigorous evolution of gas that accompanies some chemical reactions.

efflorescence—phenomenon whereby a substance evolves moisture upon exposure to the atmosphere. Also powdery exudation, usually white, on the surface of a specimen caused by precipitation or crystallization of soluble material that has migrated to the surface.

effluent—something that flows out of or forth. Commonly applied to the wastewater discharges from industrial plants, to blowdown from cooling towers, and to sanitary wastes and sewage.

eight forms of corrosion—arbitrary, but common classification of the different forms by which corrosion manifests itself. These are: (1) uniform or general corrosion; (2) galvanic or dissimilar metals corrosion; (3) crevice corrosion; (4) pitting; (5) intergranular corrosion; (6) selective leaching; (7) erosion-corrosion; and (8) stress-corrosion.

elastic deformation—change in dimensions directly proportional to and in phase with an increase or decrease in applied force. *Opposite of* **plastic deformation**.

elasticity—property of certain materials that enables them to return to their original dimensions after an applied stress has been removed.

elastic limit—maximum stress to which a material may be subjected without its permanent deformation after the stress is removed.

elastic modulus—ratio of the stress applied to a body to the strain produced.

elastohydrodynamic lubrication—form of fluid-film lubrication in which elastic deformation of the bearing surface becomes significant.

elastomer—natural or synthetic polymer that, at room temperature, can be stretched repeatedly to at least twice its original length and will immediately return to approximately its original length. Also a polymer possessing rubbery properties.

Elcometer—trademark for a magnetic direct reading dry coating film thickness gage.

electric—of, pertaining to, producing, derived from, produced, powered, or operated by electricity.

electrical conductivity—property of a fluid or solid that permits passage of an electrical current as a result of impressed electrical potential or emf.

electrical double layer—metal-solution interface where two layers of opposite charge face each other, as in a capacitor. Electric current can pass across the metal-solution interface, although there is some resistance to it. At equilibrium and during the corrosion of a metal in an aqueous solution, the metal develops an electrical potential different from that of the surrounding solution. The measured relative value is called the electrode potential. *See also* **electrode potential**.

electrical energy—form of energy related to the position of an electric charge in an electric field.

electrical resistance (ER) corrosion test method—method used to monitor the corrosion of a metal in conductive or nonconductive environments and that is based on the principle that the electrical resistance of the metal increases as its average cross-sectional area decreases due to corrosion. The electrical resistance probe utilized is simply a specially designed corrosion coupon made in convenient wire or other form such that its electrical resistance can be conveniently measured.

electrical resistivity—electrical resistance offered by a material of unit length and unit cross-sectional area or unit weight. It is the reciprocal of electrical conductivity. *Also called* **resistivity** or specific resistance.

electric arc—luminous discharge between two electrodes. The discharge raises the electrodes to incandescence with the resulting thermal ionization providing the carriers to maintain the high current between the electrodes.

electric boiler—boiler that utilizes the boiler water as resistance in order to cause heat to be generated when electrical current flows from electrode to electrode.

electric cell—*see* **cell**.

electric charge—*see* **charge**.

electric current—flow of electric charge through a conductor caused by the flow of electrons. Flow is in a direction opposite to the flow of electrons. *See also* **current**.

electric field—region in which an electric charge experiences a force usually because of a distribution of other charges.

electricity—any effect resulting from the existence and interactions of stationary or moving electric charges.

electric polarization—*see* **dielectric**.

electric potential—*see* **potential, electric**.

electric power—rate of expending energy or doing work in an electrical system.

electric spark—transient passage of an electric current through a gas between two points of high opposite potential.

electric strength—maximum potential gradient that a material can withstand without rupture. *Also called* **dielectric strength**.

electrobrightening—electrolytic process to increase the reflectivity of aluminum surfaces by removing microirregularities from the surface through an anodic smoothing. *See also* **Alzak process** and **Brytal process**.

electrochemical admittance—inverse of electrochemical impedance. *See also* **impedance**.

electrochemical cell—electrochemical system consisting of an anode and a cathode in metallic contact and immersed in an electrolyte. There are two types of electrochemical cells: (a) the galvanic cell where electrical current flows as a result of spontaneous chemical reactions proceeding at the electrodes, and (b) the electrolytic cell where the electrical current has to be impressed from an external generator to produce an effect.

electrochemical conversion coating—conversion coating produced on the surfaces of aluminum, magnesium, zinc, and titanium alloys by the anodizing process to impart corrosion and abrasion resistance. *See also* **conversion coating** and **anodizing**.

electrochemical corrosion—corrosion accompanied by a flow of electrons between cathodic and anodic areas on metallic surfaces.

electrochemical double layer—interface between an electrode and an electrolyte formed by charge-charge interaction, resulting in alignment of oppositely charged ions at the electrode surface. It has a thickness of about 10^{-7} centimeter (1 nanometer). The electrochemical double layer is also considered an electric condenser.

electrochemical equivalent—weight of an element or group of elements oxidized or reduced at 100% efficiency by the passage of a unit quantity of electricity. Usually expressed as grams per coulomb.

electrochemical method—any corrosion-monitoring method based on the electrochemical nature of corrosion.

electrochemical potential—partial derivative of the total electrochemical free energy of a constituent with respect to the number of moles of this constituent when all factors are kept constant. Also called electrochemical tension.

electrochemical potentiokinetic reactivation (EPR) test—electrochemical test to quantitatively assess the degree of sensitization of austenitic stainless steel alloys.

electrochemical reaction—chemical reaction characterized by a gain or loss of electrons at electrode surfaces.

electrochemical series—*same as* **electromotive force series**.

electrochemical tension—*See* **electrochemical potential**.

electrochemistry—chemical science and technology concerned with electrons and ions in solution reacting at a metal- or semiconductor-electrolyte interface and with interconversion of chemical and electrical energies.

electrocleaning (electrolytic cleaning)—cleaning of metal surfaces by making the workpiece an electrode in a solution of an electrolyte. Cleaning is assisted by the scrubbing action of gas bubbles liberated at the electrode. Anodic cleaning is generally preferred to cathodic cleaning because of fewer problems with side effects such as hydrogen embrittlement.

electrocoating—*see* **electrodeposition**.

electrocuring—process that uses an electron beam to cure organic coatings.

electrode—electronic conductor in contact with an ionic conductor. A conductor that emits or collects electrons. It is the interface between a metal and a solution across which a charge transfer results from an electrochemical reaction. The positive electrode is called the anode and the negative electrode the cathode. Types of electrodes that find application in electrochemistry include: (1) metal, metal ion, e.g., Cu; Cu^{2+}; (2) inert electrode, nonmetal in solution, ion, e.g., Pt; $I_2(s)$, I^-; (3) inert electrode, ions of different valence, e.g., Pt; Fe^{3+}, Fe^{2+}; (4) inert electrode, gas, ion, e.g., Pt, H_2; H^+; (5) inert electrode, neutral solutes in different states of oxidation, e.g., Pt; $C_6H_4(OH)_2$, $C_6H_4O_2$; (6) amalgam electrode, ion, e.g., (Cd + Hg); Cd^{2+}; electrode, insoluble salt or oxide, ion, e.g., Ag, AgCl; Cl^-. Any of these electrodes may be combined with any other to give a cell, the electromotive force of which is equal to the algebraic sum of the potentials of the two electrodes.

electrode kinetics—study of reaction rates at the interface between an electrode and a liquid in contact.

electrodeposition—deposition of a substance on an electrode by passing electric current through an electrolyte. *Also called* **electrocoating, electroplating,** or **plating**. Also, a process by which electrically charged paint is deposited on conductive surfaces of an opposite charge.

electrode potential—potential difference produced between the electrode and the solution in a half cell. Also the potential of an electrode measured against a reference electrode. *See also* **electrical double layer**.

electrode reaction—interfacial reaction equivalent to a transfer of charge between electronic and ionic conductors.

electrodialysis (ED)—desalting process in which ions are removed from water by passing them through a semipermeable membrane impervious to water. A direct current electrical field transports the ions through the membranes.

electrodics—electrochemical study of reduction and oxidation reactions taking place at electrodes. *See also* **ionics**.

electroendosmosis—forcing of water through a semipermeable membrane by an electrical potential in the direction of the pole with the same electrical charge as the membrane. Also refers to accelerated water diffusion through a paint coating on steel due to an electrical potential gradient established between a corroding area of the steel and the adjacent protected areas of higher electrical potential that are in electrical contact. The result is blistering.

electroforming—method of forming intricate metal articles or parts by electrodeposition of the metal on a removable conductive mold (called a mandrel or matrix).

electrogalvanizing—electrodeposition of zinc coatings.

electrogranadizing—electrolytic deposition of a coating of zinc phosphate from a zinc hydrogen phosphate electrolyte solution.

electrokinetic potential (zeta potential)—potential difference in a solution caused by residual, unbalanced charge on an adjacent surface that causes a countercharge distribution in the adjoining solution producing a double layer. It differs from the electrode potential in that it occurs exclusively in the solution phase. Electrokinetic potential causes the effects of electrophoresis and electroosmosis among other phenomena.

electroless plating—deposition of a metallic coating by a controlled chemical reduction. Also known as autocatalytic plating.

electrolysis—forced passage of electricity through an electrochemical cell that produces chemical change in the electrolyte.

electrolysis cell—container in which substances are decomposed by passing a direct electric current through them.

electrolyte—conducting liquid medium in which the flow of current is accompanied by movement of matter. A substance that exists as ions when dissolved in solution. An ionic conductor. Liquid metals are not electrolytes because conduction is by free electrons and not ions.

electrolyte solution—solution that contains ions that can result from the dissociation of certain compounds (acids, bases, or salts) in a solvent or from the melting of these at high temperature.

electrolytic—of or pertaining to electrolysis.

electrolytic capacitor—*see* **capacitor**.

electrolytic cell—electrochemical cell in which current is passed through the electrolyte from an external source.

electrolytic cleaning—*see* **electrocleaning**.

electrolytic color anodizing—development of integral or inherent colors on certain alloys by anodizing. Also known as integral color anodizing. *See also* **anodizing**.

electrolytic corrosion—corrosion that occurs through an electrochemical reaction.

electrolytic corrosion test (EC test)—polarization technique used to simulate extended atmospheric service of metal plated parts. Standardized as ASTM B 627, while the Ford accelerated corrosion test (FACT) version is standardized as ASTM B 538.

electrolytic derusting—process to remove rust where the rusted metal or alloy is made anodic in an alkaline electrolytic bath. Periodically reversing the current has also been employed.

electrolytic protection (EP)—*see* **cathodic protection**.

electrolytic refining—purification of metals by electrolysis.

electrolyze—to decompose by electrolysis.

electromagnetic force—basic force acting between electrically charged objects.

electromagnetic interference (EMI)—interaction of natural or man-made electromagnetic energy (electric or magnetic fields) with the circuitry of an electronic device. Filtering, shielding, and grounding are three means to minimize these effects and to keep electronic equipment itself from being a source of EMI to other electronics.

electromagnetic radiation—entire range of radiation extending in frequency approximately from 10^{23} cycles per second to 0 cycles per second. Includes, in order of decreasing frequency, cosmic rays, photons, gamma rays, X-rays, ultraviolet radiation, visible light, infrared radiation, microwaves, radio waves, heat, and electric currents.

electromagnetism—magnetism arising from electric charge in motion.

electrometallurgy—electrolytic processes applied to the extraction and refining of metals. Also encompasses electrolytic refining, electroplating, and the use of the electric-arc furnace.

electrometer—measuring instrument for determining a voltage difference without drawing an appreciable current from the source.

electromotive force—electrical potential difference that produces or tends to produce an electric current. Denoted as emf or, less often, EMF. Also known as electrical potential and voltage.

electromotive force of a corrosion cell—the difference in potential between the cathode and the anode. It is also a measure of the degree of thermodynamic instability of a surface metal in the given environment, i.e., the less the emf, the less the corrosion rate.

electromotive force series—list of elements arranged according to their standard electrode potentials with "noble" metals being positive and "active" metals being negative. Abbreviated as the emf series or, less often, EMF series. Also known as electromotive series and electrochemical series.

electromotive series—*same as* **electromotive force series**.

electron—subatomic particle having a negative charge.

electron beam curing—curing of paint films using the energy of an electron beam (beta rays), i.e., the high-energy radiation of the beam crosslinks the polymers.

electronegative—used to describe elements that tend to gain or have gained electrons to form negative ions. *Contrast with* **electropositive**.

electronegativity—property of atoms to attract electrons in chemical bonds. Also property of each atom that determines the direction and extent of polarity of the covalent bond that holds two atoms together.

electron energy levels—regions of space about the nucleus where electrons reside; subdivided into smaller regions called sublevels and orbitals.

electron flow—movement of electrons in an external circuit connecting an anode and cathode in a corrosion cell. The current flow is arbitrarily considered to be in an opposite direction to the electron flow.

electron gun—electron-emitting electrode and associated elements that produces a beam of accelerated electrons.

electronic—of or pertaining to electrons.

electronic configuration—arrangement and population of electrons in specific energy levels, sublevels, and orbitals in atoms.

electron micrograph—micrograph made by an electron microscope.

electron microscope—microscope that uses a beam of electrons instead of a beam of light to form a large image of a very small object.

electron pair—any two electrons functioning or regarded as functioning in concert, especially two electrons shared by two atoms joined by a covalent chemical bond.

electron probe microanalysis (EPM)—analytical method consisting of directing a very finely focused beam of electrons onto the sample to produce characteristic X-ray spectrum of the elements present. Used for elements with atomic numbers in excess of 11.

electron spin resonance (ESR)—spectroscopic method of locating electrons within molecules of a paramagnetic substance to obtain information about its bonding and structure. Method is similar to NMR (nuclear magnetic resonance) but is adapted to the much greater magnetic moment of an unpaired electron as compared with any nuclear moment.

electron volt—unit of energy equal to the energy acquired by an electron falling through a potential difference of one volt.

electroosmosis—osmotic phenomenon in which water moves through a membrane or a capillary system under the influence of an electrical potential gradient.

electrophile—ion or molecule that is electron-deficient and can accept electrons.

electrophoresis (cataphoresis)—movement of colloidal particles produced by the application of an electrical potential.

electrophoretic deposition—paint coating process by impelling electrically charged paint particles suspended in a conductive liquid toward a surface when a direct current is applied through the liquid such that the particles collect on surfaces of opposite polarity.

electrophoretic effect—slowing down, due to interionic attraction and repulsion, of the movement of an ion with its solvent molecules in the forward direction by ions of opposite charge with their solvent molecules moving in the reverse direction under an applied electrical field.

electroplating—electrodeposition, from solution, of an adherent metallic coating upon an electrode.

electropolishing—improvement in surface finish of a metal effected by making it anodic in an appropriate conductive solution. Used primarily to produce a very smooth, bright, easily cleaned surface with maximum corrosion resistance for the part.

electropositive—describes elements that tend to lose or have lost electrons to form positive ions. *Contrast with* **electronegative**.

electrorefining—purification of metals by electrolytic processes.

electrostatic—of or pertaining to stationary electric charges.

electrostatic field—electric field that surrounds a stationary charged body.

electrostatic precipitation—removal of particles suspended in a gas by electrostatic charging and subsequent precipitation onto a collector in a strong electric field.

electrostatic spraying—system of applying paint in which paint droplets or powder particles are given an electrical surface charge resulting in their attraction to a grounded workpiece.

electrothermal—of, pertaining to, or involving both electricity and heat and especially to producing heat electrically.

electrotinning—electroplating tin on an object.

electrovalence—valence characterized by the transfer of electrons from atoms of one element to atoms of another.

electrowinning—extraction of metals by electrolytic processes.

element—substance that cannot be decomposed into simpler substances by ordinary chemical means, or made by chemical union.

elimination reaction—reaction in which a molecule decomposes to two molecules, one smaller than the other.

elipsometry—irradiation of a sample surface with polarized light that is reflected and changes in the direction of polarization becomes a measure of the thickness of any surface coating.

elongation—increase in length of a material under tension; usually expressed as a percentage of the original length.

elution—process of removing an adsorbed material by washing in a liquid.

elutriation—process of suspending finely divided particles in an upward-flowing stream of air or water to wash or separate them into sized fractions.

embedded iron corrosion—potential problem during fabrication of stainless steel equipment where iron may be embedded in the stainless steel surface. The iron corrodes when exposed to moist air or when wetted, leaving rust streaks and, possibly, initiating crevice-corrosion attack in the stainless steel.

embrittlement—severe loss of ductility or toughness or both of a material.

embrittlement cracking—metal failure, predominantly intercrystalline, that occurs in steam boilers at riveted joints and at tube ends.

embrittlement-detector test—accelerated corrosion test method for determining the embrittling or nonembrittling characteristics of a water in an operating boiler. Also known as the United States Bureau of Mines (USBM) embrittlement-detector method. See ASTM D 807 for details.

emery—rock composed of corundum (natural aluminum oxide, which is Al_2O_3) with some magnetite, hematite, or spinel mineral present. Used in some grinding and polishing operations.

EMF (or emf) series—*see* **electromotive force series**.

EMI—*abbreviation for* **electromagnetic interference**; the interaction of natural and man-made electromagnetic energy (electric or magnetic fields) with the circuitry of an electronic device. Filtering, shielding, and grounding are three ways to minimize these effects and to keep electronic equipment itself from being a source of EMI to other electronics.

emission—emitting of radiation. Also discharges into the air by a pollution source as distinguished from effluents, which are discharged into water.

EMPA—abbreviation for the Swiss Federal Measuring and Testing Institute.

empirical—result obtained by experiment or observation rather than from theory.

empirical formula—representation of a chemical compound using symbols for the atoms present and giving the atoms in their simplest ratio.

emulsibility—ability of a fluid insoluble in water to form an emulsion with water.

emulsifiable oils—commonly called soluble oils, water-miscible fluids, or emulsifiable cutting fluids. They consist of oil droplets suspended in water by blending the oil with emulsifying agents and other materials. They are further subclassified as either emulsifiable mineral oils made using petroleum sulfonates or amine fatty acids, or as extreme-pressure emulsifiable oils, sometimes called heavy-duty soluble oils, that also contain sulfur, chlorine, or phosphorus additives.

emulsification—process of dispersing one liquid in another where the liquids are mutually insoluble in each other.

emulsifying agent—substance that helps to disperse one liquid in another to produce an emulsion.

emulsion—colloidal dispersion of one liquid in another.

emulsion breakers—surface-active materials and wetting agents formulated to counteract the emulsion-stabilizing agents present in raw crude oil that tend to emulsify entrained

emulsion cleaning—process for removing heavy soils from surfaces using organic solvents dispersed in an aqueous medium aided by an emulsifying agent.

emulsion paint—paint whose vehicle is an emulsion of binder in water. The binder may be oil, oleoresinous varnish, resin, or other emulsifiable binder. It differs from latex paints in which the vehicle is a latex.

emulsion polymerization—process of polymerizing a monomeric material when it is present as the discontinuous phase of an emulsion.

emulsoid—colloidal particles that take up water.

enamel—paint that is characterized by an ability to form an especially smooth film. Also a vitreous, usually opaque, protective or decorative coating baked on metal, glass, or ceramic ware. *Also called* **porcelain enamel**.

enantiotropic—element or compound able to exist in two distinct crystalline forms depending upon its maintenance above or below a certain transition temperature.

encapsulate—to conformally coat a material, component, or assembly usually by dipping, brushing, casting, or spraying.

endothermic—process in which heat flows from the surroundings to the system. Applied to reactions in which heat is absorbed. *Opposite of* **exothermic**.

endothermic occluders—metals that absorb only small amounts of hydrogen gas upon such exposure and undergo some form of hydrogen embrittlement or attack. Most common are steel, copper, silver, cerium, cobalt, and nickel. *See also* **hydride embrittlement** and **hydrogen embrittlement**.

end point—point during a titration at which the indicator changes color, indicating that the titration is completed.

energy—measure of the ability of a system to do work. Includes the following forms of mechanical energy: kinetic energy (due to motion of a body) and potential energy (due to the position or geometric configuration of a body). There are also many other forms of energy including heat, chemical, electrical, and the energy equivalences of mass. Work is energy in the process of being transferred by mechanical means from one form to another.

engine coolant—heat-exchange liquid used in the cooling system of internal combustion engines; usually a solution of water, glycol (such as ethylene or propylene glycol), and additives. It is used for removing excess heat and transferring it from the engine block and accessories to air at the radiator surface. In addition to this function of heat transfer, an engine coolant must also resist freezing in the cooling system and prevent corrosion of cooling system metals. The terms engine coolant and antifreeze are often used interchangeably. *See also* **coolant, antifreeze,** and **engine coolant concentrate.**

engine coolant concentrate—formulated liquid product intended to be diluted with water for use as an engine coolant. A mixture of glycol, usually ethylene or propylene glycol, and additives used to prepare an engine coolant. *See also* **engine coolant.**

engine dynamometer test—laboratory, full-scale engine test used to evaluate corrosion protection and corrosion inhibitors stability of engine coolants under simulated operational conditions.

engineering plastics—plastics and polymeric compositions having well-defined properties such that engineering rather than empirical methods can be used for the design and manufacture of products that require definite and predictable performance in structural applications over a substantial temperature range.

enhanced oil recovery—describes any additional oil production after primary production that results from the introduction of artificial energy into the reservoir, including waterflooding, steam flooding, and any other fluid or gas injection. Enhanced oil recovery is used for secondary and tertiary recoveries.

enthalpy—thermodynamic property of a system defined by $H = U + pV$, where H is the enthalpy, U is the internal energy of the system, p its pressure, and V its volume. Enthalpy is an energy quantity that describes the heat content of substances. The difference in enthalpy of the products and the reactants is equal to the amount of heat liberated or absorbed in a chemical reaction. For an exothermic reaction, the change in enthalpy of the system is taken to be negative in sign.

entrainment—transport of water in a gas stream.

entropy—thermodynamic probability for a chemical reaction to take place. Its symbol is S. A measure of the unavailability of energy to do work in a closed system. In a closed system, an increase in entropy is accompanied by a decrease in energy availability. In a broad sense, entropy can be interpreted as a measure of disorder, the higher the entropy the greater the disorder and therefore more energy must be put into the system to bring about its ordering.

environment—surroundings or conditions (physical, chemical, biological, mechanical) in which a material exists.

environmental cracking—brittle fracture of a normally ductile material in which the corrosive effect of the environment is a causative factor. It includes stress-corrosion cracking, hydrogen-induced cracking, liquid-metal cracking, sulfide stress cracking, corrosion fatigue, hydrogen blistering, and hydrogen- and solid-metal embrittlement.

Environmental Protection Agency (EPA)—agency of the U.S. Government that regulates and controls those agents affecting the environment.

environmental resistivity—electrical resistivity of a soil or water environment that must be known and considered when designing a cathodic protection system operative in that environment.

environmental stress cracking—*see* **environmental cracking**.

enzyme—high-molecular-weight protein produced in living things that catalyzes chemical reactions.

EPA—*abbreviation for* **Environmental Protection Agency**.

EPDM—abbreviation for a synthetic rubber that is a terpolymer made of ethylene propylene and a diene monomer.

epitaxial growth—growth of a layer of one substance on a single crystal of another such that the crystal structure in the layer is the same as that in the substrate crystal.

EPM—*abbreviation for* **electron probe microanalysis** and for **equivalents per million**.

epoxy—thermosetting polymers containing the oxirane group, i.e., where oxygen is an atom in a three-member ring. A type of paint, adhesive, or plastic noted for high mechanical strength, good adhesion, and resistance to solvents, acids, alkali, and corrosion. Most epoxies do not weather well. *See also* **epoxy resins**.

epoxy resins—synthetic resins produced by copolymerizing epoxides (compounds containing oxygen atoms in their molecules as part of a three-member ring) with phenols, e.g., epichlorohydrin and bisphenol.

EPR test—*abbreviation for* **electrochemical potentiokinetic reactivation test** method (ASTM STP 656) used to quantitatively assess the degree of sensitization of austenitic stainless steel alloys.

EPRI—abbreviation for the Electric Power Research Institute, the research arm of the electric power utility industry.

epsom salt—heptahydrated magnesium sulfate ($MgSO_4 \cdot 7H_2O$).

equilibrium—state of dynamic balance between the opposing actions, reactions, or velocities of a reversible process.

equilibrium constant—product of the concentrations (or activities) of the substances produced at equilibrium in a chemical reaction divided by the product of concentrations of the reacting substances, each concentration raised to that power which is the coefficient of the substance in the chemical equation.

equilibrium pH—pH of a water used for cooling purposes that is in equilibrium with ambient air.

equilibrium potential—potential of an electrode in an electrolyte when the forward rate of a given reaction is exactly equal to the reverse rate. The equilibrium potential can only be defined with respect to a specific electrochemical reaction. Also known as the reversible potential.

equivalence point—point in an acid-base titration when the number of moles of H^+ from the acid equals the number of moles of OH^- from the base.

equivalent conductance—conductance of a volume of solution of an electrolyte containing one equivalent weight of dissolved substance when placed between two parallel electrodes 1 centimeter apart, and large enough to contain between them all of the solution.

equivalent conductivity—conductivity of a solution containing one equivalent of a given salt.

equivalents per million (EPM)—concentration in parts per million of any ion divided by its equivalent weight.

equivalent weight (combining weight)—weight in grams of an element or compound that could combine with or displace one gram of hydrogen (or 8 grams of oxygen or 35.5 grams of chlorine) in a chemical reaction. For an element, it is obtained by dividing the atomic weight by the valency. For a compound, it depends on the reaction considered.

erosion—wearing away of material due to mechanical interaction between that surface and a fluid or with solid particles. *See also* **wear**.

erosion-corrosion—conjoint action of corrosion and erosion in the presence of a moving corrosive fluid. Process whereby a flowing fluid surface destroys a solid surface and where corrosion exacerbates the destruction.

erosive wear—any of four distinct forms of wear: solid particle impingement, slurry erosion, liquid droplet impingement, and cavitation erosion.

error—*see under statistical terms*.

ESCA—abbreviation for electron spectroscopy used for chemical analysis. *See also* **X-ray photoelectron spectroscopy**.

ESR—*abbreviation for* **electron spin resonance**.

essential amino acids—amino acids that are not biosynthesized; amino acids that must be obtained through dietary sources.

essential element—any element required by living organisms to ensure normal growth.

esters—organic compounds formed by reaction between alcohols and acids. Esters formed from carboxylic acids have the general formula *RCOOR′*.

estuary—part of the wide lower course of a river where its current is met and influenced by the tides.

etch—to dissolve or mechanically abrade unevenly a part of the surface of a material.

etch primer—*see* **wash primer**.

ethanolamine—organic compound of composition $NH_2(CH_2)_2OH$. Also known as monoethanolamine (MEA).

ether—class of organic compounds that contains two hydrocarbon groups bonded to an oxygen atom (*R–O–R′*).

eutectic—solid solution of two or more substances that solidify as a whole when cooled from a liquid state without change in composition. Also a mixture of two or more substances that has the lowest melting point.

eutectoid—phase transformation where a single solid phase is transformed into two or more different solid phases.

eutrophic—natural body of water with abundant supply of nutrients and a high rate of formation of organic matter by photosynthesis. *Contrast with* **oligotrophic**.

eutrophication—enrichment of water causing excessive growth of aquatic plants and eventual choking and deoxygenation of the water body.

Evans diagram—kinetic diagram representing the electrode potential in volts on the ordinate and the electrochemical reaction rate in amperes on the abscissa.

evaporation—change of state of a liquid into a vapor at a temperature below the boiling point of the liquid. Evaporation occurs at the surface of a liquid.

evaporation rate—rate at which water is evaporated to cool a circulating water. Usually expressed as a given volume per minute.

evaporative cooling—cooling of water by exposing its surface to air. The heat transfer involves chiefly a latent heat transfer due to evaporization and a lesser transfer due to the difference in temperature between the water and air.

evaporators—devices to convert water to pure vapor that is condensed leaving saline and other solute residues behind. Used for such procedures as preparing boiler feed water and evaporating seawater to produce fresh water.

exchange current—rate at which either positive or negative charges are entering or leaving the surface of an electrode in dynamic equilibrium in a solution such that the rate of anodic dissolution balances the rate of cathodic plating.

exchange current density—rate of electron exchange between the two phases of an electrode when the system is in equilibrium.

excursion—cooling tower term. Any change in cooling water characteristics that takes it outside of control limits.

exempt solvents—solvents whose use is not subject to air pollution legislation. Includes many alcohols, esters, and some ketones.

exergonic reaction—reaction that releases free energy to the surroundings.

exfoliate—scaling from a surface in flakes or layers.

exfoliation corrosion—corrosion that proceeds laterally from the sites of initiation along planes parallel to the surface and generally at grain boundaries to form corrosion products that force metal away from the body of the material, giving rise to a layered appearance. Exfoliation is a structure-dependent form of localized corrosion, usually intergranular, most familiar in certain alloys and tempers of aluminum.

exfoliation tests—accelerated laboratory tests for the susceptibility of aluminum specimens to exfoliation corrosion include the Annexes A2, A3, and A4 of ASTM G 85, and also ASTM G 66, and ASTM G 34.

exothermic—process in which heat flows from the system to the surroundings; applied to reactions in which heat is liberated. *Opposite of* **endothermic**.

exothermic occluders—*see* **hydride embrittlement**.

exposure tests—tests conducted to evaluate the durability of a coating or film by exposure to sunlight or ultraviolet light, moisture, cold, heat, salt water, mildew, etc.

extenders—inert pigments that, however, can contribute specific properties to paint or plastic. They are usually lower in cost than the primary opacifying, active, or colorant pigments. Extenders are often used as a filler or for economy.

exterior basecoat—coating applied to the outside of such products as beverage cans to provide corrosion resistance and/or a lithographic or printing surface.

exterior paints and varnishes—coatings formulated for use in conditions exposed to the weather and expected to possess reasonable durability.

external circuit—wires, connectors, measuring devices, current sources, etc., that are used to bring about or measure the desired electrical conditions within a test cell.

extractive metallurgy—study and practice of the removal of metals from minerals.

extreme-pressure lubrication (EP lubrication)—lubrication where special extreme pressure or boundary lubrication agents are utilized either by themselves or as additives to a grease or oil in order to resist the negative effects of contact between surfaces. The agents most commonly employed are reactive sulfur, chlorine, or phosphorus compositions used as soluble additives and molybdenum disulfide or graphite used as solids.

extrusion (of metals)—forming technique in which steady compression forces on unheated metal cause its controlled flow into a die.

extrusion (of plastics)—forming technique for plastics that involves their being compacted and forced through an orifice.

exudation—migration of a substance to the surface.

F

face rusting—appearance of rust on an otherwise apparently unblemished painted surface.

FACT test—abbreviation for the Ford anodized aluminum corrosion test (ASTM B 538), which is a rapid test for quality of anodized aluminum in automotive applications.

facultative organisms—microbes capable of adapting to either aerobic or anaerobic environments.

fadeometer—accelerated test apparatus for determining the resistance of resins and paint coatings to fading by subjecting the material to high-intensity ultraviolet wavelengths similar to those found in natural sunlight.

fading—subjective term to describe the lightening of the color of a material following its exposure to the effects of light, heat, time, temperature, chemicals, etc.

Fahrenheit—designation of a degree on the Fahrenheit temperature scale that is related to the International Practical Temperature Scale by means of the equation: $T_F = 1.8\ T_C + 32$, where T_F is the temperature in degrees Fahrenheit and T_C is the temperature in degrees Celsius.

failure analysis—examination of materials after their failure to determine cause and, hopefully, to develop a remedy. Accomplished by using any or all of visual and microscopic examination, chemical analysis, physical tests, and other appropriate procedures.

farad—SI unit of capacitance, being the capacitance of a capacitor that, if charged with one coulomb, has a potential difference of one volt between its plates. Symbol is F.

Faradaic efficiency—*see* **current efficiency**.

faraday—quantity of electricity that is capable of depositing or dissolving one gram equivalent weight of a substance in electrolysis, approximately 96,500 coulombs. A quantity of electric charge equal to 96,500 coulombs.

Faraday cage—ideal enclosure for an electronic device to completely eliminate electromagnetic interference. The enclosure completely surrounds the device and is usually made of metal wire.

Faraday's laws—two laws describing electrolysis: (1) the amount of chemical change during electrolysis is proportional to the charge passed; (2) the charge required to deposit or liberate a mass m is given by $Q = Fmz/M$, where F is the Faraday constant, z the charge of the ion, and M is the relative ionic mass.

fastness—ability of a pigmented or dyed material to resist color changes following exposure to light or the environment.

fat—esters of higher fatty acids and glycerine. Fats can be made of animal or vegetable materials or can be made synthetically.

fatigue—process leading to fracture or decrease in strength of a material due to repetitive loading well below the normal tensile strength of the material.

fatigue strength—maximum stress that can be sustained by a material for a specific number of stress cycles without failure due to fatigue.

fatigue wear—wear of a solid surface caused by fracture arising from material fatigue.

fatty acid—organic compound consisting of a hydrocarbon chain of from 6 to 20 carbons and a terminal carboxyl group (–COOH). Originally used to describe those organic acids derived from fats and fatty oils.

fatty oil—fat that is liquid at normal room temperature.

FDA—abbreviation for the Food and Drug Administration.

FEP—*abbreviation for* **fluorinated ethylene propylene**.

fermentation—anaerobic degradation of carbohydrates, especially to ethanol and CO_2, by various microorganisms. Also called souring.

ferric—iron in its +3 oxidation state.

ferric chloride test—procedure for determining the relative resistance of alloys to crevice corrosion in an oxidizing acidic chloride solution. See ASTM G 48 for details.

ferric ion corrosion—attack of ferrous metals by the combination of aqueous acidity and ferric ions and that is further accelerated by atmospheric air. Most commonly associated with acid mine waters. Also a form of general corrosion occurring on carbon steel, nickel, Monel 400, copper, brass, and zirconium during chemical cleaning with mineral acids. Results when deposits of iron oxide dissolve during cleaning and in the presence of aeration generate the oxidizing Fe^{3+} ion.

ferric oxide—Fe_2O_3; stable anhydrous oxide of iron and a constituent of mill scale.

ferrite—solid solution of one or more elements in body-centered cubic iron. Also alpha iron containing alloying elements in solid solution. Also a member of a class of mixed oxides of the general formula $MO \cdot Fe_2O_3$ where M is a metal such as cobalt, manganese, nickel, or zinc; the oxides show either ferri- or ferromagnetism but are not electrical conductors.

ferritic stainless steels—magnetic stainless steels containing about 14 to 18% chromium, a maximum of 0.5% nickel, less than 0.12% carbon and no other major alloying elements except that some grades may contain up to 4% molybdenum for improved pitting and crevice corrosion resistance. They are characterized by their ferritic structure, and they are not hardenable by heat treatment, only by cold work. Used principally for their good corrosion resistance and high temperature properties. They are part of the "400" series of stainless steels, the most common being Type 405, 430, and 436.

ferroalloys—alloys of iron with other metals made by smelting mixtures of iron ore and the other metal ore. They are used in making alloy steels.

ferrous—iron-containing metal or alloy as contrasted with nonferrous. Also of, pertaining to, or containing iron. Also iron in its +2 oxidation state.

ferroxyl test—test for porosity of nickel coatings on steel.

FGD—*abbreviation for* **flue gas desulfurization**.

FHSA—abbreviation for the Federal Hazardous Substances Act.

FHWA—abbreviation for the Federal Highway Administration.

fiberglass—glass in fibrous form. Individual filament or group of filaments made by attenuating molten glass.

fiberscope—flexible instrument containing two fiberoptic bundles used for visual inspection of otherwise inaccessible surfaces.

field coat—coat of paint applied at the site of erection or fabrication.

FILC—*abbreviation for* **flow-induced localized corrosion**.

filiform corrosion—random small threads of corrosion that develop beneath some semipermeable coatings and films. The essential conditions for its development generally are a high humidity (65 to 95% relative humidity at room temperature), sufficient water permeability of the film, stimulation by impurities, and the presence of film defects (mechanical damage, pores, insufficient coverage of localized areas, air bubbles, salt crystals, or dust particles).

fill and soak cleaning—chemical cleaning method for relatively small parts or equipment involving filling the equipment or soaking the part with the cleaner and draining it after a set period of time; may be repeated several times.

filler—solid, inert material added to a synthetic resin or rubber either to change its physical properties or simply to dilute it for economy.

filliform—slender as a thread.

film—thin surface layer that may or may not be visible to the naked eye.

film-former—substance that forms a skin or membrane when dried from a liquid state.

filming amines—chemicals added to condensate of boiler systems to reduce both oxygen and carbon dioxide corrosion by replacing the loose oxide scale on the metal surfaces of the system with a very thin protective amine film. A commonly used filming amine is octadecylamine, which is used alone or blended with emulsifiers and neutralizing amines for better coverage and protection. *See also* **neutralizing amines**.

film integrity—continuity of a coating free of defects.

filmogen—film-forming material.

filter—device for separating solid particles from a liquid or gas.

filter cake—suspended solids deposited on a porous medium during the process of filtration.

filter paper—porous, unsized paper used to filter liquids for purpose of separation of any entrained insoluble matter.

filtrate—liquid remaining after removal of solids as a cake in a filter or filter paper.

filtration—process of separating solids from a liquid by means of a porous substance through which only the liquid passes.

fines—portion of a powder whose particles are smaller than a specified size.

fingernail test—gouging a dried film with the fingernail to make a subjective, qualitative estimate of the hardness and toughness of the film.

fingerprint neutralizer—*see* **fingerprint removers** and **rust preventive compounds**.

fingerprint removers—solvent-type coatings that serve to clean fingerprint residues from metallic surfaces on application and on drying to act as temporary corrosion protectives. They generally are low-viscosity compositions containing suitable solvents to dissolve, suppress, and neutralize acids, salts, and residues from handling. The coatings are removed with a solvent before the application of a subsequent longer-term rust protection coating. Fingerprints are body excretion residues consisting primarily of urea, salts, acids, water, and natural body oils.

fire—destructive burning as manifested by light, flame, heat, and/or smoke.

fire and smoke corrosion—corrosion of metals exposed to fire and smoke where complex interactions occur between combustion products and the normally protective oxide films formed by metal alloy systems.

fire scale—black oxidation product of iron formed during the heat treatment of iron or steel composed of an inner layer of ferrous oxide (FeO), accompanied by a thinner layer of magnetic iron oxide (Fe_3O_4) and a top, very thin layer of ferric oxide (Fe_2O_3). Also called hammer scale.

fireside corrosion—corrosion of boiler tubes brought on by the combustion of fuels producing gases, vapors, and particulates that create aggressive chemical environments.

fireside deposits—potentially corrosive deposits resulting from the burning of certain fuels producing slags.

firing—controlled heat treatment of ceramic materials in a kiln or furnace to develop the desired properties.

fish oil—general designation of natural oils extracted from fish and characterized by their large content of saturated fatty acids commonly associated with mixed triglycerides.

fish paper—electrical-insulation grade of vulcanized fiber in thin cross section.

fissure—surface split or crack.

fixed matter—residues from the ignition of particulate or dissolved matter, or both.

Flade potential—pH-dependent potential at which the corrosion current for a given metal falls to the passive value corresponding to the onset of full passivity. Also called the passivation potential.

flake pigments—pigments consisting of flat particles.

flame cleaning—method for the surface preparation of metals using flame to burn off contaminants, remove mill scale, and dehydrate any remaining rust to leave the surface suitable for wire brushing and the subsequent application of paint.

flame descaling—method of removing millscale by heating the scale layer to crack it off due to differential expansion.

flame-retarding agents—material used as a coating or on a component of a combustible product to raise its ignition point. Agents are classified as nondurable (water-soluble inorganic salts easily removed by washing or exposure to water), semidurable (removed by repeated washing), and durable (not affected by laundering).

flame spraying—thermal spraying in which a coating material is fed into an oxyfuel gas flame where it is melted and propelled onto a substrate to form a coating.

flammable—subject to easy ignition and rapid combustion. *See also* **inflammable**.

flammable liquid—any liquid having a flashpoint below 37.8 °C (100 °F) and having a vapor pressure not exceeding 40 psi (absolute) at 37.8 °C. Excludes any mixture having components with flashpoints of 37.8 °C or higher, the total of which make up 99% or more of the total volume of the mixture (OSHA definition).

flash—portion of a superheated fluid converted to vapor when its pressure is reduced.

flashoff—the escape of dissolved substances into the atmosphere due to forced-air stripping.

flash plate—thin electrodeposit that is less than 0.1 mil thick.

flashpoint—minimum temperature at which a liquid gives vapor within a test vessel in sufficient concentration to form an ignitable mixture with air near the surface of the liquid (OSHA definition).

flash rusting—very thin film of rust occurring on a surface within minutes to several hours after exposure to moisture. Also the onset of rusting on a paint film after applying a wet film of certain waterborne coatings.

flaw—imperfection in an item or material.

fleet test—*see* **vehicle service test**.

flexibilizer—inert or a reactive material added to rigid plastics to make them resilient or flexible. Sometimes called plasticizer, but a flexibilizer is more specifically restrictive in function than a plasticizer.

flexural strength—property of a material indicating its ability to withstand a flexural or transverse load.

floc—loose, open-structured mass produced in a suspension by the aggregation of minute particles.

flocculate—to aggregate into larger particles or masses.

flotation-type coatings—rust-retarders applied to the inside surfaces of tanks by depositing a layer of inhibited oil on the bottom of empty tanks through being introduced from the bottom so that the oil film floats on top and as the water level rises and falls, a layer of the coating is deposited.

flow cavitation—cavitation caused by a decrease in static pressure induced by changes in velocity of a flowing liquid.

flowcoating—system of applying paint in which the paint is allowed to flow over and drain off the workpiece.

flow-induced corrosion—conjoint action of flow and corrosion. Four types of corrosion are generally recognized as being flow-induced: (1) mass transport-controlled corrosion, (2) phase transport-controlled corrosion, (3) erosion-corrosion, and (4) cavitation corrosion.

flow-induced localized corrosion (FILC)—corrosion failure mechanism in fast-flowing media mostly produced at flow perturbations such as at steps, grooves, and surface imperfections like corrosion pits, weld beads, tiny obstacles, corrosion product scales, etc.

flue gas desulfurization (FGD)—cleaning application for flue gases. Typically accomplished using wet scrubbing units with lime or limestone slurries for sulfur dioxide absorption.

flue gases—gases that pass through a flue and that are commonly classified as oxidizing, i.e., contain an excess of oxygen along with carbon dioxide, water vapor, and possibly sulfur dioxide, or reducing, i.e., contain lesser amounts of the oxidizing gases found in oxidizing flue gases and include carbon monoxide, hydrogen, and hydrogen sulfide.

fluid—gas, liquid, or any substance that can be caused to flow. Substance that readily assumes the shape of the container in which it is placed.

fluid friction—friction due to the viscosity of fluids.

fluidity—property of substances to flow. The reciprocal of viscosity. *Opposite of* **viscosity**.

fluidization—procedure in which solid particles suspended in a stream of gas are treated as though they were in the liquid state.

fluidized bed coating—process of applying a coating in which a heated or electrostatically charged article is immersed or passed over a fluidized bed of powdered coating such that a coating is formed that is subsequently adhered by the hot metal or by catalyzed curing.

fluidizers—chemical additives that distort crystal structures so that crystals do not stick to surfaces or to each other.

fluorescence—property of emitting radiation as the result of absorption of radiation from some other source. The emitted radiation persists only as long as the exposure is subjected to radiation, which may be electrified particles or waves.

fluorescent dye—dye that fluoresces, i.e., gives off visible light, when it is exposed to short-wavelength radiation such as ultraviolet or near-ultraviolet light.

fluorescent penetrant—penetrant incorporating a fluorescent dye to improve the visibility of a flaw.

fluorescent UV lamp—lamp in which radiation from a low-pressure mercury arc is transformed to longer wavelength UV by a phosphor of characteristic spectrum.

fluorinated ethylene propylene (FEP)—copolymer of tetrafluoroethylene and hexafluoropropylene having a service temperature of about 175 °C (350 °F) with corrosion resistance similar to that of PTFE.

fluorocarbon—organic compounds analogous to hydrocarbons where the hydrogen atoms have been replaced by fluorine.

fluorocarbon plastic—plastic based on polymers made with monomers composed of fluorine and carbon only.

flushing—passing fresh water through heat exchangers to remove loose particles, dirt, and debris.

flux—substance applied to the surfaces of metals to be soldered in order to inhibit their oxidation or corrosion. Also a substance used in the smelting of metals to assist in the removal of impurities as slag. Also a substance that promotes fusion in a given ceramic mixture.

fly ash—finely divided particles of ash, usually 100 microns or less, entrained in flue gases arising from the combustion of coal.

FMA—*abbreviation for* **free mineral acidity**.

foam—dispersion of gas bubbles in a liquid or a solid. Solid foams are made by foaming a liquid or pasty phase of a material then allowing it to set, cure, or otherwise harden while retaining the foam structure.

fog—loose term applied to visible aerosols in which the dispersed phase is liquid. Also a dispersion of water or ice in the atmosphere.

FOG—abbreviation for fats, oil, and grease.

fogged metal—metal whose luster has been reduced due to the formation of a thin corrosion product layer.

fogging—phenomenon occurring in cooling-tower operation that produces a highly visible plume and possible icing hazards. Results from mixing warm, highly saturated tower

force

discharge air with cooler ambient air that lacks the capacity to absorb all the moisture as vapor.

force—agency that tends to change the momentum of a body defined as being proportional to the rate of increase of momentum. It is a vector quantity that tends to produce an acceleration of a body in the direction of its application.

forced draft aeration—aeration of water accomplished by using spray nozzles, packing, or wood slat fill to break-up water into small droplets or into a thin film to enhance countercurrent air contact.

forced drying—acceleration of drying by increasing the drying temperature above ambient using such means as an oven, infrared lamp, or other heat source.

foreign matter—insoluble foreign particles in or on a material.

foreign structure—any associated metallic structure not intended as part of a cathodic protection system.

formaldehyde—HCHO. Colorless, pungent, gas, readily soluble in water and used as a disinfectant and catalyst and as a hardener for certain synthetic resins.

formalin—37% by weight aqueous solution of formaldehyde (HCHO).

formation damage—damage to the productivity of a oil or gas well resulting from invasion of mud particles or mud filtrates into the formation.

formic acid—HCOOH. Also called methanoic acid or hydrogen carboxylic acid. Used in electroplating for pH and redox adjustment, as a food preservative, and germicide.

form oil—oil used to lubricate wooden or metal concrete forms to keep cement from sticking to them.

formula—(1) way of representing a chemical compound using symbols for the atoms present and subscript numbers for the number of each atom present. (2) list of materials and their amounts used in the preparation of a compound or mixture.

formula weight—relative molecular mass of a compound as calculated from its molecular formula. Its weight in grams, pounds, or other units, obtained by adding the atomic weights of all elemental constituents in a chemical formula.

fossil—remains or traces of any organism that lived in the geological past.

fossil fuels—natural materials burned directly as fuels. Includes natural gas, crude oil, peat, coal, or wood. Also includes the by-product fuels obtained by processing natural carbonaceous materials (diesel fuel, gasoline, kerosene, and synfuels such as producer gas,

fouling—term used to describe the action by which submerged surfaces are covered by sessile marine growth, such as barnacles. Also, the deposition of a thermally insulating material onto a heat-transfer surface. In the case of fouling by cooling water, the deposit may consist of one or more of crystalline compounds, amorphous inorganic materials, silt, corrosion products, or biological growths. Fouling deposits from organic liquids are principally coking or polymerization products.

fouling organism—any aquatic organism that attaches to and fouls an underwater structure.

foul water—*same as* **sour water**.

Fourdrinier—papermaking machine for producing paper in a continuous roll or web.

four-pin method—*see* **Wenner method**.

FPM (fpm)—abbreviation for feet per minute.

fps units—British system of units based on the foot, pound, and second. Now replaced for all scientific purposes by SI units.

fractional crystallization—separation of a mixture of soluble solids by dissolving them in a suitable hot solvent then lowering the temperature slowly to where the least soluble component will crystallize out first, followed sequentially by the others.

fractional distillation—separation of a liquid into its different components by collecting fractions of distillates over a restricted boiling range.

fractography—descriptive treatment of metallic fracture using photographs of the fracture surface.

fracture mechanics—quantitative analysis for evaluating structural behavior of materials in terms of applied stress, crack length, and specimen geometry.

fracture strength—normal stress at the beginning of fracture.

fracture toughness—resistance to extension of a crack.

free available chlorine—*see* **chlorine, free available**.

freeboard—exposed side of the ship hull between the upper waterline and the main deck.

free carbon—that part of the total carbon in steel or cast iron that is present in elemental form as graphite or temper carbon.

free corrosion potential—corrosion potential in the absence of net electrical current flowing to or from the metal surface.

free energy—thermodynamic property expressing the resultant enthalpy of a substance and its inherent probability (entropy). Its symbol is G. It is the energy liberated or absorbed in a reversible process at constant temperature and constant pressure. If the change in free energy is positive, a reaction will only occur if energy is supplied to force it away from the equilibrium condition. If the change is negative, the reaction will proceed spontaneously. Free energy is a measure of the ability of a system to do work.

freely corroding potential—*see* **corrosion potential**.

free mineral acidity (FMA)—hydrogen ion concentration (pH) attributed to mineral acids (sulfuric, hydrochloric, nitric, etc.). Measured by titration with a base from a lower pH value up to pH 4.3.

free radical—atom or group of atoms with an unpaired valence electron. Also a very reactive neutral chemical species having the aforementioned characteristic.

freeze—to pass from the liquid to the solid state by loss of heat. Also to acquire a coat of ice from cold, to turn rigid and inflexible, solidify.

freeze-drying—process for removing moisture from a wet material by bringing the material to the solid state and subsequently subliming it. Used primarily for drying and preserving food products.

freezing point—temperature at which a liquid and solid are in equilibrium. The same temperature as the melting point.

freezing point depression—decrease in the freezing point of a solvent after the addition of a solute. *See also* **colligative property**.

frequency—number of cycles per second of alternating electric current.

fresh water—either surface or ground water, and typically containing less than 1% sodium chloride.

fretting (fretting wear)—form of wear resulting from an oscillating or vibratory motion of limited amplitude that results in the removal of very finely divided particles from rubbing surfaces.

fretting corrosion—deterioration at the interface between contacting surfaces due to the conjoint action of fretting and corrosion. *See also* **wear corrosion**.

fretting fatigue—reduction in fatigue life of a part subjected to fretting and operating under fatigue loading.

friction—resisting force encountered at the common boundary between two bodies when, under the action of an external force, one body moves or tends to move relative to the surface of the other.

friction coefficient—ratio of the force required to move one surface over another to the total force pressing the two together.

friction oxidation—*see* **fretting corrosion**.

frit—product made by quenching and breaking up a glass. Frits are the basic material of porcelain enamel coating.

frost—deposit or covering of minute ice crystals formed from frozen water vapor.

FRP—abbreviation for fiberglass or other fiber-reinforced plastic. Terms used interchangeably with FRP are RTP (reinforced thermoset plastic), RTR (reinforced thermoset resin), and GRP (glass-reinforced plastic).

FSCT—abbreviation for the Federation of Societies for Coatings Technology.

FTIR—abbreviation for Fourier transform infrared spectroscopy.

FUCA—abbreviation for the fluorescent UV and condensation apparatus used for rapidly evaluating the physical deterioration of paint coatings (by cracking, checking, etc.) and the fastness of plastics to light.

fuel—substance oxidized or otherwise changed in a furnace or heat engine to release useful heat or energy.

fuel-ash corrosion—corrosion of incinerators, boilers, heat exchangers, gas turbines, calciners, and recuperators by molten fuel-ash deposits.

fuel/ballast tank—tank used to carry ship's fuel but that can be filled with seawater as the fuel is used thus preventing unbalancing of the ship. The fuel floats on the seawater.

fuel cell—cell in which the chemical energy of a fuel is converted directly into electrical energy. Also an electrochemical device that directly combines hydrogen and oxygen from air to produce electricity and water.

fugacity—symbol f. The equivalent pressure of a real gas for which the ideal gas equations are valid.

fuller's earth—nonplastic variety of kaolin clay.

fume—solid particles generated by condensation from the gaseous state. Also gaslike emanation containing minute solid particles arising from the heating of a solid body. Differs thus from a gas or vapor.

functional group—group of atoms responsible for the characteristic reactions of a compound.

functional group analysis—analytical determination of specific organic groups and radicals in organic matter.

fundamental units—set of independently defined units of measurement forming the basis of a system of units.

fungi—unicellular or filamentous nucleated organisms that do not contain chlorophyll. Fungi utilize carbohydrates that are synthesized by green plants. They grow on nonliving organic matter, or as parasites on other living organisms. Two types are commonly found in cooling water systems: molds and yeasts. Molds cause either white or brown rot of the cooling-tower wood depending on whether they are cellulolytic (attack cellulose) or lignin-degrading. Yeasts can produce slime and are also cellulolytic. *See also* **fungus**.

fungicide—substance poisonous to fungi.

fungistat—agent that inhibits the germination of fungus spores or the development of mycelium.

fungus—any of a group of plants such as molds, yeast, mildew, mushrooms, smuts, etc.

fuse—to melt or join together by melting.

fused quartz—*see* **vitreous silica**.

fusion—melting.

fusion coating—powder coating that melts, fuses, and reacts chemically as it contacts a heated surface.

fusion, nuclear—nuclear reaction in which atomic nuclei of low atomic number fuse to form a heavier nucleus with the release of large amounts of energy.

FVV—abbreviation for the Forschungsvereinigung Verbrennungskraftmaschinen (Internal Combustion Engines Research Association), which is a federation of German car manufacturers.

G

galena—lead sulfide (PbS). Used for increasing the density of oil and gas well drilling fluids to points impractical or impossible with baryte.

galling—form of wear in which seizing or tearing of the surface occurs.

Galvalume—trademark for precoated sheet steel subjected to a hot dip galvanizing process using a zinc alloy to produce a coating consisting approximately of 55% aluminum, 1.5% silicon, and 43.5% zinc.

galvanic anode—metal that, because of its relative position in the galvanic series, provides sacrificial protection to metals that are more noble in the series when both are coupled in an electrolyte.

galvanic cell—electrochemical cell with its electrode reactions proceeding spontaneously. Also, an electrolytic cell capable of producing electrical energy by electrochemical action.

galvanic corrosion—corrosion of a metal because of its electrical contact with a more noble metal while in the presence of an electrolyte. *Also known as* **bimetallic corrosion** and displacement corrosion. *See also* **couple**.

galvanic couple—pair of dissimilar metallic conductors in electrical contact.

galvanic current—electric current flowing between metals or between conductive materials in a galvanic couple.

galvanic pain—dentistry term describing the discomfort arising from contacting dissimilar-alloy restorations. The electrochemical circuit is short-circuited by contact, giving an instantaneous current flow through the external circuit, which is the oral tissues.

galvanic protection—reduction or elimination of corrosion of a metal achieved by making current flow to it from a solution by connecting it to the negative pole of some source of current. The source of the protective current for steel is a sacrificial metal such as zinc, magnesium, or aluminum.

Galvanic Series—list of metals and alloys arranged according to their relative corrosion potentials in a given environment.

galvanized—pertaining to zinc-coated articles.

galvanized iron—iron or steel that has been coated with a layer of zinc to protect it from corrosion.

galvanizing—application of a coating of zinc. Also called zinc hot dipping when produced by immersion in a bath of molten zinc.

galvanneal—zinc-iron alloy coating on iron or steel developed by keeping a coating molten after hot dip galvanizing until the zinc alloys completely with the base metal so that the coating contains approximately 10 to 12% iron.

galvanometer—instrument for detecting and measuring small electric currents.

galvanostatic—experimental technique whereby an electrode is maintained at a constant current in an electrolyte.

gamma iron—face-centered cubic form of pure iron stable from 910 to 1400 °C (1670 to 2550 °F).

gamma rays—quanta of electromagnetic wave energy similar to but of much higher energy than ordinary X-rays. Gamma rays have no mass or charge. *See also* **electromagnetic radiation**.

gangue—rock or other waste material present in an ore.

garnet—group of silicate minerals used as abrasives that conform to the general formula $A_3B_2(SiO_4)_3$, where A may include magnesium, calcium, manganese, and iron(II), and B may include aluminum, iron(III), chromium, or titanium.

gas—state of matter in which the matter concerned occupies the whole of its container irrespective of its quantity. An ideal gas obeys the gas laws exactly. Its molecules have negligible volume and negligible forces between them and their collisions are perfectly elastic. Real gases deviate from the gas laws because their molecules occupy finite volume, there are small forces between its molecules, and their collisions are to a certain extent inelastic.

gas-cap-drive oil reservoirs—oil field recovery process. It is used in reservoirs where a cap of gas occurs over the oil reservoir. As reservoir pressure due to its dissolved gas becomes lowered and dissolved-gas-drive recovery becomes diminished, the gas cap expands to help fill the pore spaces formerly occupied by oil thus yielding greater recovery of oil. *See also* **dissolved-gas-drive oil reservoirs**, **water-drive reservoirs**, and **oil-well pumps**.

gas chromatography (GC)—technique for quantitatively separating or analyzing a mixture of organic compounds. The vaporized sample and gas move and are partitioned between two phases, an inert moving gas and a stationary liquid film in a long column. The separation depends on the affinity each component has for the stationary phase. Compounds with the least affinity emerge from the column first and others later. Separately, they pass through a detector and each presence is indicated by a peak on a recorder. Comparison with standards determines identity and concentration.

gas constant (R)—numerical proportionality constant of volume, pressure, temperature, and moles in the ideal gas equation, $pV = nRT$, and whose numerical value is 0.082056 L · atm/(K · mol), where p is the absolute pressure, V is the volume, n is the number of moles in the gas sample, and T is the temperature in kelvin.

gas cyaniding—misnomer for carbonitriding.

gaseous corrosion—corrosion reaction with gas as the only corrosive agent and without any aqueous phase present. Also called dry corrosion. *Contrast with* **wet corrosion**.

gaseous cutting fluids—typically, air used in dry cutting operations, but also includes use of other gases such as argon, helium, and nitrogen where oxidation must be prevented.

gasket corrosion—specific example of crevice corrosion occurring with a gasketing material.

gas-liquid chromatography (GLC)—technique for separating or analyzing mixtures of gases by chromatography. In GLC, a liquid is used as the stationary phase to act as solvent for the sample components. *See also* **chromatography**.

gasoline—mixture of hydrocarbons composed mainly of alkanes having formulas ranging from C_5H_{12} to $C_{12}H_{26}$ and of many isomeric forms. Gasoline also contains small amounts of other kinds of hydrocarbons and some sulfur- and nitrogen-containing compounds.

gas phase corrosion—another term for "reducing-atmosphere corrosion" occurring in coal- and oil-fired boilers resulting from direct reaction of the water wall tubes with a substoichiometric gaseous environment containing sulfur or with deposited, partially combusted char containing iron pyrites. The reducing conditions have two main effects on corrosion. They tend to lower the melting temperature of any deposited slag increasing its ability to dissolve normal protective oxide scale on the tubes, and the stable gaseous sulfur compounds, which include H_2S, react to form iron sulfide. The latter scale allows significantly higher rates of transport of iron cations than oxides do and so are less protective.

gas-solid chromatography (GSC)—technique for separating or analyzing mixtures of gases by chromatography. In GSC, a solid material is used as the absorbent or stationary phase. *See also* **chromatography**.

gastric juice—acidic fluid inside the stomach that contains aqueous HCl, inorganic salts, and digestive enzymes (e.g., pepsin).

gate—electronic circuit with a single output that is a function of one or more inputs.

gauss—Symbol G. The c.g.s. unit of magnetic flux density and equal to 10^{-4} tesla.

Gay-Lussac's law—*see* **Charles' law**.

GC—abbreviation for **gas chromatography**.

Geiger counter—device used to detect and measure ionizing radiation.

gel—semisolid system consisting of a network of solid aggregates in which liquid is held.

gelatin—hot-water-soluble protein type material extracted from animal skins, sinews, tendons, and bones that has gelling characteristics.

gelation—formation of a gel from a liquid state.

gel coat—thin outer layer of resin applied to the surface of a mold used for reinforced plastic molding. It is gelled prior to lay-up and becomes part of the finished laminate. It is usually used to improve surface appearance of the reinforced plastic molding.

gel-time—reaction rate of a given thermoset powder from the time it melts until it becomes a semisolid.

general corrosion—corrosion occurring in a more or less uniform manner over the surface of a material. *Also called* **uniform corrosion.**

generator—machine that converts mechanical power into electrical power.

generic—belonging to a particular family. Also nonproprietary.

geothermal energy—thermal energy contained in the rocks and fluids of the earth that is a potential source of useful energy. Volcanoes, geysers, hot springs, and fumaroles are all sources of geothermal energy.

germ—microorganism, especially a pathogen, from which a new organism may develop.

germicide—chemical designed to kill all bacteria present. Any agent that kills germs. Germicides may be classified as (1) oxidants such as peroxides generating atomic oxygen and chlorine-, bromine-, and iodine-yielding products; (2) quaternary ammonium compounds; (3) phenolics such as carbolic acid and substituted phenolics such as cresylic acids and chlorinated phenols; (4) anilides such as tribromo- or trichlorocarbanilides; (5) pine oil; (6) metal compounds such as tributyl tin, mercury phenyl acetate, bismuth, zinc, copper and silver compounds; and (7) gases such as formaldehyde, glutaraldehyde, ethylene oxide, and triethylene glycol.

Gibbs free energy—maximum useful work that can be obtained from a chemical system without a net change in temperature or pressure.

gilding—art of covering surfaces with layers of gold leaf.

gilsonite—asphalt found in Utah. One of the purest of natural bitumens.

glacial acetic acid—pure acetic acid, CH_3COOH.

glass—inorganic product of fusion that has cooled to a rigid condition without crystallizing. It may be transparent, translucent, or opaque, and it may be colored.

glass coatings—includes vitreous enamel, glass linings, or porcelain enamels as coatings on metallic substrates.

glass electrode—type of half cell having a glass bulb containing an acidic solution of fixed pH into which dips a platinum wire. The glass bulb is thin enough for hydrogen ions to diffuse through it. Used for pH measurement.

glass transition temperature—temperature at which a noncrystalline polymer is transformed from a rubbery material to a brittle, glasslike material.

glassware corrosion test—laboratory screening test for evaluating the corrosion susceptibility of a material under controlled environmental conditions.

glaze—hard, glassy, fused coating.

GLC—*abbreviation for* **gas-liquid chromatography**.

gloss—sheen, brightness, luster, or the ability to reflect light from a dried surface.

glue—adhesive prepared from the hides, tendons, cartilage, bones, etc., of animals by heating them with water.

glue line—thin layer of adhesive that attaches two adherends.

glycol—organic molecule that contains two hydroxyl (–OH) groups.

glycol engine coolant concentrate—engine coolant concentrate that is subsequently diluted with water for use in an engine as a coolant and has the property of a depressed freezing point and elevated boiling point. The chief component of the concentrate is usually ethylene glycol, which usually also contains dissolved chemicals such as chemical corrosion inhibitors to protect the metallic components of the cooling system from corrosion and antifoaming agents to inhibit foaming while in use. *See also* **antifreeze**.

gouging abrasion—severe form of abrasive wear in which the force between an abrading body and the wearing surface is sufficiently large that a macroscopic gouge, groove, deep scratch, or indentation can be produced in a single contact.

gpg—abbreviation for grains per gallon; $gpg \times 17.1 = ppm$.

gpm—abbreviation for gallons per minute.

graft polymer—a high polymer, the molecules of which consist of two or more polymeric parts of different composition and chemically united. *See also* **copolymer**.

grain—individual crystal in a polycrystalline matrix.

grain boundary—interface separating two grains whose crystal axes are differently oriented.

grain-boundary corrosion—*see* **intergranular corrosion**.

grain dropping—dislodgement and loss of a grain or grains from a metal surface as a result of intergranular corrosion.

grains per gallon—concentration measurement of calcium carbonate hardness of water. One grain per U.S. gallon is equal to 17.1 parts per million $CaCO_3$. One grain per imperial gallon is equal to 14.3 parts per million $CaCO_3$.

grain refiner—additive to molten metal prior to its casting and used to obtain finer grains in the casting.

grainy—having a uniformly roughened surface detectable by touch.

gram—one-thousandth of a kilogram. Symbol is g.

gram atomic weight—mass or weight in grams numerically equal to the atomic weight.

gram equivalent weight—weight in grams of a substance involved in an oxidation-reduction reaction that is equivalent to one mole of electrons.

gram molecular weight—amount of a pure substance having a weight in grams numerically equal to its molecular weight.

granite—visibly granular, igneous rock consisting mostly of quartz and feldspars and usually accompanied by dark-colored minerals.

graphite—type of elemental carbon, characterized by its layer-lattice crystallites.

graphitic corrosion—corrosion of gray cast iron (2 to 4% carbon) where the metallic constituents are selectively leached or converted to corrosion products leaving the graphite intact. *See also* **graphitization**.

graphitization—formation of graphite in iron or steel from decomposition of iron carbide at elevated temperatures. Also a substitute term for graphitic corrosion. *See also* **graphitic corrosion**.

gravel—coarse, granular aggregate with pieces larger than sand grains and resulting from the natural erosion of rock.

gravimetric analysis—type of quantitative analysis that depends on weighing of sample, precipitate, etc., as the underlying basis of calculation.

gray blast—abrasion cleaning term equivalent to commercial blast clean. *See also* **commercial blast**.

gray iron—cast iron with a large proportion of the graphitic carbon present in the form of flake graphite.

gray water—wastewater of a system that may also include water-carried solid or insoluble liquid wastes except for human wastes.

grease—lubricant composed of an oil thickened with a soap or other thickener to a semi-solid, pastelike, or solid consistency. Also any lubricating agent of higher viscosity than oils.

grease paint—nondrying and nonoxidizing coating used in the void spaces of ships, usually for corrosion protection.

greenhouse effect—effect occurring in the atmosphere because of the presence of certain gases that absorb infrared radiation leading to an increase in the temperature of the earth and its atmosphere.

green liquor—liquor resulting from dissolving smelt from the kraft recovery furnace in water. *See also* **smelt**.

green plague—green corrosion deposits formed in copper hot-water piping and on water faucets thought to be due to some effect of electrical grounding on copper dissolution.

green rot—high-temperature corrosion of chromium-bearing alloys where green chromium oxide forms while other alloy constituents remain metallic.

green rust—greenish corrosion product on ferrous metals that contains iron in two oxidation states and has a variable anion content that may include OH^-, Cl^-, BO_2^-, CO_3^{2-}, or SO_4^{2-}. Its type formula is $[Fe_4^{2+}Fe_2^{3+}(OH)_{12}]^{2+}(X^-, X^{2-})_2 \cdot 3H_2O$. Structurally related to the pyroaurite group of naturally occurring minerals, and its structure consists of cationic layers, $Fe(OH)_2^{n+}$, alternating with anionic layers stacked in a six-layer repeating sequence.

grinding—removal of small chips of metal from the workpiece by the mechanical action of irregularly shaped grains imbedded into a grinding wheel.

grit blasting—cleaning of surfaces by abrasive blasting with small irregular pieces of steel, malleable cast iron, aluminum oxide, or any other crushed or irregularly shaped abrasive. *See also* **abrasive blasting**.

grit number—synonym for mesh number.

grooving corrosion—localized corrosion, manifested as grooves, occurring in the weld of electric resistance welded carbon steel pipe exposed to aggressive waters. Caused by the redistribution of sulfide inclusions along the weld line during the welding process.

groundbed—buried item, such as junk steel or graphite rods, that serves as the anode for the cathodic protection of pipelines or other buried structures. Also the ground connection in a cathodic protection system.

ground state—electrons of an atom at their lowest energy level and that may be excited when energized by some external source to emit energy, usually as light.

ground water—that part of the subsurface water that is in the saturated zone, i.e., water that percolates into the earth's crust and collects in subterranean pools and underground rivers. Well water.

ground wire—wire attached to an object to dissipate electrostatic charge.

grout—gypsum or Portland cement plaster used to fill crevices or hollow metal frames.

GSA—abbreviation for the General Services Administration.

GSC—*abbreviation for* **gas-solid chromatography**.

gum—any of a class of colloidal substances exuded by or prepared from plants and that is composed of complex carbohydrates and organic acids which are soluble or swell in water and exhibit stickiness when moist.

Gunite—trade name for sprayed-on concrete.

gun metal—bronze alloy usually containing 88 to 90% copper, 8 to 10% tin, and 2 to 4% zinc. Exhibits high resistance to wear and corrosion and is commonly used as a bearing material.

gypsum—mineral form of $CaSO_4 \cdot 2H_2O$. Used in wallboard, industrial plasters, Portland cement, and agriculturally as a soil conditioner.

gypsum cement—plaster of Paris and calcined gypsum.

H

hackles—thin, needlelike protrusions found on steel plates that have been blasted with steel or grit.

halide attack—corrosion of metals or alloys by halogens, principally chlorine and fluorine, which oxidize the metal atoms to form a halide layer scale.

half cell—electrode immersed in a suitable electrolyte forming part of a cell. A half cell can also be formed by a metal in contact with an insoluble salt or oxide and an electrolyte, e.g., calomel half cell.

halide—compound of a halogen with another element or group.

halite—naturally occurring sodium chloride (NaCl). *See also* **rock salt**.

halogen—any element of the halogen family, i.e., fluorine, chlorine, bromine, iodine, and astatine.

hammer scale—*see* **fire scale**.

hand lance—long steel tube with a jet nozzle on the end used to apply high-pressure water to the entire internal length of a heat-exchanger tube to be cleaned.

HAP—abbreviation for Hazardous Air Pollutants of the Clean Air Acts of 1977 and 1990 that identify 189 chemicals as hazardous.

hard anodizing—anodizing for engineering rather than decorative purposes. Most processes employ a refrigerated electrolyte at below 10 °C (50 °F), vigorous agitation, and close control of processing conditions.

hard asphalt—matter formed by the oxidation of mineral oil and that is insoluble in petroleum spirit.

hardboard—smooth, grainless panel manufactured from wood fibers consolidated under heat and pressure and to which other materials may have been added for property modification.

hard coating—development of an anodic oxide coating on aluminum with a higher apparent density and thickness and a greater resistance to wear than a conventional anodic coating. Also known as hardcoat anodize.

hard chromium—chromium electroplated for engineering rather than decorative purposes.

hardener—curing agent; promoter; or catalyst for polymers.

hardening—heating and quenching certain ferrous alloys from either within or above the critical range to produce a hardness superior to that than if not quenched.

hard facing—thermal deposition of filler material on a metal surface to improve wear resistance. The filler material adheres to the base metal by fusion or metallurgical bonding.

hard lead—lead alloy containing about 4 to 15% antimony that is used where chemical lead lacks the necessary strength. *See also* **chemical lead**.

hardness—(1) degree to which a material will withstand pressure without deformation or scratching. (2) term used to describe the polyvalent cation concentration of water (generally its calcium and magnesium content). Typical U.S. units for reporting water hardness are parts per million as $CaCO_3$, or grains $CaCO_3$ per U.S. gallon. One grain is equivalent to 17.1 ppm. The Degree French is used in France. One Degree French = parts per 100,000 as $CaCO_3$ (=10 ppm U.S.). The Degree Clark is British. One Degree Clark = grains per imperial gallon as $CaCO_3$ (=14.3 ppm U.S.). The Degree German is used in Germany. One Degree German = part per 100,000 calculated as CaO and not $CaCO_3$ (=17.8 ppm U.S.).

hard radiation—ionizing radiation of high penetrating power, usually refers to gamma rays, or short-wavelength X-rays.

hard water—water containing certain salts, such as those of calcium and magnesium, that form insoluble deposits in boilers and precipitate with soap. *See also* **temporary hardness** and **permanent hardness**.

hardwoods—trees having broad leaves (deciduous) in contrast to conifers or softwoods, but having no reference to the hardness of the wood. *See also* **softwoods**.

Hastelloy—trademark for group of nickel alloys particularly resistant to corrosion by alkaline solutions. Typically comprised of nickel and chromium, molybdenum, carbon, and small amounts of other elements.

Hastelloy B—trademark for nickel alloy containing 30% molybdenum and 5% iron.

Hastelloy C—trademark for nickel alloy containing 15% chromium, 16% molybdenum, 4% tungsten, and 5% iron.

Hazard Communication Standard (Worker Right-to-Know Law)—regulation requiring all workers be made aware of the hazards associated with their work.

hazardous substance—substance that, by reason of being explosive, flammable, poisonous, corrosive, oxidizing, or otherwise harmful, is likely to cause death or injury when misused.

hazardous waste—any waste material that can cause or contribute to death or illness, or that threatens human health or the environment when improperly managed.

HDPE—*abbreviation for* **high-density polyethylene**.

heartwood—inner layer of a woody stem wholly composed of nonliving cells. The wood extending from the pith to the sapwood. More resistant to decay than sapwood.

heat—form of kinetic energy that when transferred to an object increases its temperature if the object is not undergoing a state change.

heat-affected zone—portion of the base metal not melted during brazing, cutting, or welding, but whose microstructure and properties were altered by the heat.

heat capacity—quantity of heat required to increase the temperature of a system or substance one degree of temperature (usually expressed in calories per degree centigrade).

heat check—pattern of parallel surface cracks on a metal surface formed by alternate rapid heating and cooling.

heat exchanger—device for transferring heat from one fluid to another without permitting the two fluids to contact each other.

heat load—heat removed from a circulating cooling water and expressed as calories or British thermal units (Btu) per unit time.

heat of combustion—energy liberated when one mole of a given substance is completely oxidized.

heat of formation—energy liberated or absorbed when one mole of a compound is formed from its constituent elements. The energy released as heat when a metal reacts with oxygen in the air and that can also be expressed as the electrical potential between a metal and an aqueous solution of a standard concentration of the ions of the metal.

heat of reaction—energy liberated or absorbed as a result of the complete chemical reaction of molar amounts of the reactants.

heat pump—device for transferring heat from a low-temperature source to a high-temperature region by doing work.

heat stabilizers—compounds added to resins to prevent their thermal degradation during molding, extrusion, or later use when exposed to heat. Most commonly they are lead and organotin compounds, or mixed metal-salt blends based on barium, cadmium, and zinc.

heat transfer—transfer of energy in the form of heat from one body or system to another as a result of a difference in temperature. There are three generally accepted methods for transferring heat from one medium to another, or from one locale to another, conduction, convection, and radiation.

heat treatment—heating and cooling of a metal or alloy in the solid state to obtain certain desirable conditions or properties.

heavy duty oil—oil having oxidation stability, bearing corrosion preventive properties, and detergent dispersant characteristics making it suitable for use in both high-speed diesel and gasoline engines under heavy-duty service. Also a mineral oil containing oxidation inhibitors, detergent, and other additives, and suitable for heavy-duty diesel engines.

heavy metal—metal with a relatively high atomic mass; usually applied to the common transition metals. In the case of industrial wastewaters, heavy metals is the classification generally applied to copper, silver, zinc, cadmium, mercury, lead, chromium, iron, and nickel content.

heavy phosphates—coarse crystalline phosphate conversion coatings that contain divalent metal ions from solution and from the metal surface with coating weights ranging from 7.5 to 30 g/m^2 (700 to 2800 mg/ft^2) and exhibiting increased unpainted corrosion resistance. They include manganese phosphates, zinc phosphates, and ferrous phosphates.

heavy water—water in which hydrogen atoms, ^1H, are replaced by the heavier isotope deuterium, ^2H. Also called deuterium oxide. *See also* **light water**.

hematite—iron mineral and common ore of iron. Chemically, it is Fe_2O_3. Also the iron oxide component of scale on steel that exhibits one of a lower solubility rate in pickling acids than do magnetite and wustite iron oxides.

hemihydrate—hydrated compound containing two molecules of compound per molecule of water, e.g., $2CaSO_4 \cdot H_2O$.

hemipolymers—readily soluble polymers of molecular weights between about 1000 and 10,000.

Henry's law—at a constant temperature, the mass of gas dissolved in a liquid at equilibrium is proportional to the partial pressure of the gas.

Herbert test—laboratory test to assess the corrosivity of aqueous solutions or dispersions, used as cutting fluids, toward cast iron. In this test, steel millings are placed on the cleaned surface of a cast iron plate and the fluid under test is poured onto them. After 24 hours, the millings are removed and the surface of the plate examined for corrosion. Standardized as the IP 125 method.

hertz—symbol Hz; SI unit of cyclic frequency and equal to one cycle per second.

hetero atom—odd atom in the ring of a heterocyclic compound.

heterocyclic compound—cyclic compound composed of molecules that have more than one type of atom in the ring; usually a cyclic organic compound that contains nitrogen, oxygen, or sulfur in the ring.

heterogeneity—having different properties at different points. *Contrast with* **homogeneity**.

heterogeneous alloys—alloy mixtures of two or more phases that are separate because their components are not completely soluble.

heterogeneous mixture—mixture composed of two or more distinct components. Usually applied to mixtures with more than one observable phase.

HEW—abbreviation for the Department of Health, Education, and Welfare.

hiding power—ability to obscure a substrate.

high-boiling-point engine coolant—mixture of glycol engine coolant concentrate and water. *See also* **low-boiling-point engine coolant**.

high boiling solvent—solvent with an initial boiling point above 150 °C (302 °F).

high build—term referring to producing thick (minimum 5 mils) dry films per coat.

high build paints—paints containing a structuring or gelling agent that allows thick films to be applied without sagging.

high density polyethylene (HDPE)—linear polyethylene plastic having a density of 0.941 g/cm^3 or greater.

high flash solvent naphtha—paint solvent consisting of aromatic hydrocarbons related to xylene. Its flashpoint is about 43 °C (110 °F).

high-performance liquid chromatography (HPLC)—chromatographic technique in which the sample is forced through the chromatography column under pressure. *See also* **chromatography**.

high polymer—general term for a polymer of high molecular weight.

high-purity water—water containing very little solids, gases, and exhibiting a very high electrical resistance (above 200,000 Ohm-cm and usually about 2 megohms).

high-silicon cast iron—term generally referring to highly corrosion resistant cast iron having a silicon content of more than 14%.

high-solids (HS) coatings—paint coatings laid down from formulations having a reduced volatile content compared with conventional paints.

high-solids paints—paints containing 50% or more solids by volume. *Also called* **high solids coatings**.

high-speed steel—steel alloy that will remain hard at dull red heat and can therefore be used in cutting tools for high-speed lathes. It usually contains 12 to 22% tungsten, up to 5% chromium, and 0.4 to 0.7% carbon along with small amounts of other metals.

high-temperature corrosion (high-temperature oxidation)—metallic corrosion resulting from exposure to oxidizing gases at elevated temperature by direct reaction without need for the presence of an electrolyte. Also refers to high-temperature tarnishing, high-temperature oxidation, and high-temperature scaling. The oxidation process may form oxides, sulfides, or carbides. In hydrogen, at elevated temperatures and pressures, the availability of atomic hydrogen can penetrate metal to react internally with reducible species, e.g., carbon or cuprous oxide to form methane and water, respectively, leading to fissures or voids.

high-temperature hydrogen attack—loss of strength and ductility of steel due to the high-temperature reaction of absorbed hydrogen with carbides in the steel resulting in decarburization and internal fissuring.

high-tensile steel—term used for the high carbon steel normally used in concrete prestressed products.

hindered settling—settling stage where the accumulated settled solids compact sufficiently to hinder their water displacement and hence further settling is slowed.

HIS—abbreviation for harmonic impedance spectroscopy, an electrochemical procedure for monitoring instantaneous corrosion rates.

histogram—vertical bar chart of the frequency distribution of data.

HLB—*abbreviation for* **hydrophile-lipophile balance**.

hoar frost—frozen dew that forms a white coating on a surface. Also called white frost. *See also* **frost**.

holding time index (HTI)—expression of the half-life of a treatment chemical added to an evaporative cooling water system.

hold paint—marine paint for the holds of ships.

holiday—discontinuity in a coating; either porosity, crack, or a gap.

homogeneity—relating to only one phase. A description of uniformity of distribution, chemical composition, or physical property. *Contrast with* **heterogeneity**.

homogeneous alloy—alloy that is a solid solution, i.e., its components are completely soluble in one another and the material has only one phase.

homogeneous mixture—mixture of substances that generally has the same composition throughout. Also a solution.

homogenizing—process for reducing the size of particles in a liquid. A reduction of globule size in a mixture of two immiscible liquids makes an emulsion possible.

homopolymer—polymer formed from a single monomer.

honing—abrasive machining process designed to improve bore geometry and surface finish.

hot ash corrosion—accelerated corrosion found in boiler tubes or on gas turbine blades operating at high temperatures in contact with combustion gases of crude oils high in vanadium. The cause is generally thought to be the formation of a low-melting oxide phase that acts as a flux to dislodge or dissolve any protective Fe_3O_4 scale. The oxidation products tend to be voluminous and porous. Hot ash corrosion is a specific example of catastrophic oxidation.

hot corrosion—corrosion of metal resulting from the combined effect of oxidation and reaction with sulfur compounds or other contaminants, such as chlorides, to form a molten salt on the metal surface.

hot cracking—weldment cracking caused by segregation at the grain boundaries of low-melting constituents in the weld metal.

hot dip coating—metallic coating developed by dipping the basis metal into a molten metal. Usually applies to coatings of zinc (galvanizing), aluminum (aluminizing), or lead and tin (terne coating).

hot melt adhesive—thermoplastic adhesive composition usually solid at room temperature that is heated to a fluid state for application.

hot melt coatings—compositions that liquefy readily on heating and are applied to various surfaces in molten condition. Usually involve coal tar or asphalt.

hot rolled steel—steel that is hot reduced, i.e., formed and shaped while hot.

hot spray—spraying material heated to reduce its viscosity.

hot strippers—process for chemical paint stripping. Normally associated with the use of hot aqueous alkali solutions. Typically formulated with sodium hydroxide, chelating agents, surface activating agents, and, possibly, solvents other than water.

hot-water pitting—electrochemical pitting of copper in domestic hot water systems and associated generally with microdeposited cathodic materials such as manganese dioxide, hydrated hematite, and aluminum hydroxide as a result of copper losing its protective film.

hot working—deforming metal plastically at such a temperature and strain rate that recrystallization takes place simultaneously with the deformation thus avoiding any strain hardening. *Contrast with* **cold working**.

house paint—paint designed for use on large exterior surfaces of a building.

HPLC—*abbreviation for* **high-performance liquid chromatography**.

HRB—abbreviation for the Highway Research Board.

HS—*abbreviation for* **high solids**.

HTHW—abbreviation for high-temperature hot water.

HTI—*abbreviation for* **holding time index**.

Huey test for stainless steels—laboratory test where the alloy is exposed to boiling 65% nitric acid for five 48-hour periods and the corrosion in mils per year is reported. Details of test are given in ASTM A 262.

Hull cell—trapezoidal box of nonconducting material with electrodes arranged to permit observation of cathodic or anodic effects over a wide range of current densities.

hull paint—bottom paint used on hulls of ships.

humectant—moistening agent. Substance that has an affinity for water with a stabilizing action on the water thereby promoting moisture retention.

humic acid—organic acids of indefinite composition in naturally occurring leonardite lignite.

humidify—to increase, by any process, the quantity of water vapor within a given space.

humidistat—regulatory device used for the automatic control of relative humidity and activated by changes in humidity.

humidity—condition of the atmosphere with respect to its water vapor content.

humidity, absolute—*see* **absolute humidity**.

humidity ratio—weight of water vapor per unit weight of dry air.

humidity, relative—*see* **relative humidity**.

humification—process by which organic matter decomposes.

humus—dark-colored, amorphous colloidal material that constitutes the organic component of soil.

HVAC—abbreviation for heating, ventilating, and air conditioning.

hydrate—compound in which molecules of water, as such, are present. The water may be present as "cationic water," as in coordination compounds, where it is apparently joined to cations by covalent bonds; as "anionic water," where it is joined to anions through covalent bonds, or more frequently, through hydrogen bonds; as "lattice water," where water molecules occupy definite positions in the crystal lattice but is not coordinated with either cations or anions; as water molecules in holes of lattices; or in essentially noncrystalline materials such as hydrous precipitates and colloidal gels.

hydrated lime—dry powder obtained by treating quicklime with enough water to satisfy its chemical affinity for water and consisting essentially of calcium hydroxide or a mixture of calcium hydroxide and magnesium hydroxide.

hydration—act of a substance to take up water by means of absorption or adsorption.

hydraulic cement—cement that is capable of setting and hardening under water due to interaction of water and constituents of the cement.

hydraulic cleaning—mechanical cleaning method using high-pressure water at from 300 to 32,000 psig to flush and scour mud and debris from pipelines. Method is also used to remove light rust, minerals, and polymers from surfaces. Periodic air injection aids the scouring action.

hydraulic fluid—liquid used in hydraulic systems for transmitting power.

hydraulic lime—lime containing more than 10% silicates and that will harden under water.

hydraulic spraying—spraying by using hydraulic pressure. *Same as* **airless spraying**.

hydrazine—chemical compound N_2H_4; a powerful reducing agent used as a scavenger of dissolved oxygen in water, mainly boiler water and the cooling water for nuclear reactors. It reacts with oxygen forming nitrogen and water, and in the absence of oxygen in boiler water it acts as a sink for dissolved oxygen that may enter later. Hydrazine may

hydride—ionic or covalent binary compound of hydrogen. Examples of ionic hydrides are LiH and CaH_2, and of covalent hydrides NH_3 and SiH_4.

hydride embrittlement—form of hydrogen embrittlement characterized by hydride phase formation with certain metals that absorb large quantities of hydrogen and are known as exothermic occluders (vanadium, titanium, zirconium, tantalum and thorium). *See also* **hydrogen embrittlement**.

hydroblasting—cleaning with high-pressure water jet.

hydrocarbon—organic compound consisting of only hydrogen and carbon.

hydrodynamic lubrication (fluid-film lubrication)—fluid lubrication where the lubricant film is sufficiently thick to prevent the opposing surfaces from coming into contact.

hydrogen-assisted stress cracking—results when the metallic component is stressed in a hydrogen or hydrogen sulfide environment. *Also called* **hydrogen embrittlement**.

hydrogenation—addition reaction in which hydrogen adds to an unsaturated organic compound.

hydrogen attack—reaction of hydrogen with readily reducible carbides or oxides within an alloy to form methane or steam impairing the structural integrity of the alloy. A high-temperature attack where the hydrogen source is usually from some process gas. It is not a form of hydrogen embrittlement.

hydrogen blistering—formation of blisterlike bulges on a ductile metal surface due to internal hydrogen pressure formed during cleaning, plating, or corrosion.

hydrogen bond—type of electrostatic interaction between electronegative atoms in one molecule and hydrogen atoms bound to electronegative atoms in another molecule.

hydrogen cycle—operation of a cation-exchange cycle wherein the removal of specified cations from the influent water is accomplished by exchange with an equivalent amount of hydrogen ions from the exchange material.

hydrogen damage—general term describing various mechanical damage of a metal caused by the presence of, or interaction with, hydrogen. It is a form of environmentally assisted failure most often resulting from the combined action of hydrogen and residual or applied stress. Includes hydrogen environment embrittlement, hydrogen stress cracking, hydrogen attack, hydrogen blistering, hydride formation, hydrogen microperforation, hydrogen flow degradation, and decarburization.

Note: The top of the page begins with "also decompose in the boiler forming ammonia, hydrogen, and nitrogen with the ammonia providing pH control in the condensate."

hydrogen disease—specific rupture damage, along grain boundaries, to copper containing free cuprous oxide (Cu_2O), resulting from its tendency to dissolve oxygen when heated in air. Caused by formation of steam when the metal is subsequently heated in hydrogen. It is a form of internal oxidation.

hydrogen disintegration—deep internal cracks in a metal caused by hydrogen.

hydrogen electrode—type of half cell in which a metal foil is immersed in a solution of hydrogen ions and hydrogen gas is bubbled over the foil. The standard hydrogen electrode uses a platinum foil in a one molar solution of hydrogen ions with the gas at one atmosphere pressure and the half cell at 25 °C (77 °F). The hydrogen electrode is also called a hydrogen half cell. *See also* **hydrogen scale**.

hydrogen embrittlement—embrittlement of a metal or alloy caused by absorption of hydrogen. It may occur during any operation where the metal reacts with water to provide a source for hydrogen and most commonly during melting, casting, hot working, welding, pickling, cathodic cleaning, or electroplating. *Also called* hydrogen-induced cracking, **hydrogen-assisted stress cracking**, hydrogen-assisted stress-corrosion cracking, and **hydrogen stress cracking**.

hydrogen environment embrittlement—form of hydrogen embrittlement resulting during the plastic deformation of alloys in contact with hydrogen-bearing gases or a corrosion reaction. It is strain-rate dependent, being greatest when the strain rate is low and hydrogen pressure and purity are high.

hydrogen equivalent—number of replaceable hydrogen atoms in one molecule or the number of atoms of hydrogen with which one molecule of a substance could react.

hydrogen half cell—*see* **hydrogen electrode**.

hydrogen-induced cracking (HIC)—*see* **hydrogen embrittlement** and **hydrogen stress cracking**.

hydrogen ion concentration—measure of the acidity or alkalinity of a solution. It is normally expressed as pH. *See also* **pH**.

hydrogen overvoltage—overvoltage associated with the liberation of hydrogen. *See also* **overvoltage** and **overpotential**.

hydrogen scale—arbitrary zero electrode potential assigned to a reversible hydrogen electrode. See hydrogen electrode.

hydrogen stress cracking—spontaneous brittle fracture of highly stressed high-strength steel components contacted by sour water or sour gas (i.e., hydrogen sulfide). Also called hydrogen-induced cracking, static fatigue, and hydrogen-assisted cracking. *See also* **hydrogen embrittlement**.

hydrogen test probe—instrument used to anticipate the potential for hydrogen damage (blistering, embrittlement, stress-corrosion cracking) by a metal or alloy tube based on the diffusion rate of hydrogen through it.

hydrogeology—science of the properties, distribution, and circulation of subsurface water.

hydrological cycle—endless circulation of water between the earth and the atmosphere.

hydrolysis—chemical reaction of a compound with water. Also the reaction of a salt with water to form an acid and base. In coatings technology, it refers to a decomposition process or paint failure due to reaction with water.

hydrolytic stability—ability to withstand environmental effects of high humidity.

hydrometer—floating instrument for determining the specific gravity of a liquid, solution, or slurry.

hydronium ion—ion resulting when a hydrogen ion combines with water; $H^+ + H_2O = H_3O^+$.

hydrophilic—having a strong affinity for water, i.e., wettable. *Opposite of* **hydrophobic** and **lipophilic**.

hydrophile-lipophile balance (HLB) value—empirical value of arbitrary units of 0 to 20 based on the premise that all surfactants combine both hydrophilic and lipophilic groups on one molecule and that the weight percentages of these two groups is an indication of the extent of their behavior, either hydrophilic or lipophilic.

hydrophobic—tending to repel water. *Opposite of* **hydrophilic**.

hydrosol—sol in which the continuous phase is water. *See also* **colloid**.

hydrosphere—water on the surface of the earth that includes the oceans, seas, rivers, and lakes. 74% of the earth's surface is covered with water; 97% of which is in the oceans and seas as seawater, 2% is fresh water locked in ice caps and glaciers, and 1% resides as fresh water in rivers, lakes, and the atmosphere.

hydroxide—metallic compound containing the ion OH^- (hydroxide ion), or containing the group $-OH$ (hydroxyl group) bound to a metal atom.

hydroxyl—chemical radical $-OH$ in organic compounds and having a basic nature.

hygrometer—any instrument for measuring the humidity in an atmosphere.

hygroscopic—substances capable of absorbing water from the atmosphere.

hyper- —prefix denoting over, above, high, more than, etc.

hyperbolic cooling towers—natural draft cooling towers built of concrete and whose height and basin width are equal, a shape allowing for optimum air flow and structural stability.

hypo- —prefix denoting under, below, low, etc.

hypochlorites—salts of hypochlorous acid (HOCl). The sodium and calcium salts are added to water systems to function as disinfectants much in the same manner as chlorine gas.

hyposulfite—*see* **sulfite**.

I

IACS—abbreviation for the International Annealed Copper Standard, which sets annealed copper as having 100% electrical conductivity as a basis against which other metals, alloys, and materials are compared. Only silver (108.4%) exceeds the conductivity of annealed copper. Aluminum is rated at about 61%, steel 11%, and nickel-chromium alloy at 1.5%.

ice point—temperature at which there is equilibrium between ice and water at standard atmospheric pressure, i.e., the freezing or melting point of ice under standard conditions.

ICP—abbreviation for inductively coupled plasma emission spectroscopy, an analytical technique for the determination of metals in a sample.

ideal gas—hypothetical gas that obeys the gas laws, which relate the temperature, pressure, and volume of such a gas. Essentially, the ideal gas consists of molecules occupying negligible space, having negligible forces between them, and whose collisions are perfectly elastic. Also called perfect gas.

IEC—abbreviation for the International Electrotechnical Commission.

ignition—initiation of combustion.

ignition temperature—temperature to which a substance must be heated before it will burn.

imbibition—uptake of water by substances that do not dissolve in water so that the process results in swelling of the substance.

immersion coatings—organic coatings for marine service that resist moisture absorption, moisture transfer, and electroendosmosis (electrochemically induced diffusion of moisture through the coating) in addition to being strong and adherent. Cathodic protection is usually used to supplement the protection afforded by the immersion coatings. The most common coatings for immersion service are coal tar epoxies and straight epoxies.

immersion corrosion testing—laboratory test method where a metal or alloy is immersed in a solution or liquid medium and its weight loss measured over a given period of time.

immersion plating—process for producing a metallic deposit by a displacement reaction (which is also a controlled corrosion reaction) in which one metal displaces another from solution. Also known as chemical deposition.

immiscible—descriptive of two or more fluids that are mutually insoluble.

immunity—state of resistance to corrosion or anodic dissolution in a particular environment caused by the thermodynamic stability of the metal.

impact plating—*see* **mechanical plating**.

impact strength—amount of energy required to fracture a material under an impact load.

impedance—quantity that measures the opposition of a circuit to the passage of a current and therefore determines the amplitude of the current. Also called electrochemical impedance. *See also* **electrochemical admittance**.

impedance test—procedure used to evaluate the seal quality of anodized aluminum (ASTM B 457).

imperial gallon—English gallon equal to 1.2 U.S. gallons.

impingement—process of a succession of impacts between liquid or solid particles and a solid surface.

impingement attack—localized erosion-corrosion resulting from the turbulent or impinging flow of liquids. Also called impingement corrosion.

impressed current—electric current supplied by a device employing a power source external to the electrode system, e.g., direct current for cathodic protection. *See also* **impressed current anode**.

impressed current anode—application of cathodic protection using a ground bed anode that is not dependent upon an external source of electrical energy and, instead, uses some external source of direct current power connected (impressed) between the structure to be protected and the ground bed anodes.

impressed current cell—corrosion cell where the electrical current is supplied from an outside source, e.g., an electroplating cell.

inactive cutting oils—cutting oils of mineral oil basis generally compounded with chemically inactive additives providing high lubricity, but exhibiting limited antiweld properties. They are classified further as being either straight mineral oils for very light duty operations, compounded cutting oils with polar additives and/or chemically active additives, fatty mineral oils, which are formulated blends of straight mineral oil with up to 40% fatty oils, and inactive extreme-pressure cutting oils, which contain sulfur, chlorine, or phosphorus additives but are formulated only for light duty use.

incineration—consumption by burning.

inclusion—any foreign matter in a body that is held mechanically.

incompatibility—inability to mix with or adhere to another material.

Inconel—trademark for a group of nickel alloys.

Inconel 600—trademark for nickel alloy containing 16% chromium and 7% iron noted for its resistance to oxidizing aqueous media.

incubation period—time period prior to the detection of corrosion in which the metal or alloy is in contact with the corrodent.

indicator—substance, which by its color, shows the presence of a chemical substance or ion. Most commonly they are acid-base indicators exhibiting characteristic color at a specific pH or a range of colors over a range of pH, or oxidation-reduction indicators showing a color change in the presence of an oxidant or reductant species.

inductance—property of an electric circuit or component that causes an emf to be generated in it as a result of a change in the current flowing through the circuit or of a change in the current flowing through a neighboring circuit with which it is magnetically linked.

industrial atmosphere—atmosphere in an area of heavy industry with soot, fly ash, and sulfur compounds as the principal contaminants.

industrial coatings (or finishes)—coatings applied to factory-made articles either before or after their fabrication. *Contrast with* **trade sales paints** and **architectural coatings**.

industrial environment—*same as* **industrial atmosphere**.

industrial hygiene—laboratory and production practice of materials handling that recognizes chemical hazards due to exposure of human tissue to chemicals and avoids both the short- and long-term hazards.

industrial maintenance paints—high-performance coatings formulated to give resistance to heavy abrasion, water immersion, chemicals, corrosion, temperature, electrical current, or solvents.

industrial waste—water-conveyed residues resulting from manufacturing or processing operations.

industrial wastewater—discharge water from an industrial process and resulting from its formation or utilization in that process.

industrial water—water used in industrial processes.

inert—not participating in any fashion in chemical reactions.

inert anode—anode that is insoluble in the electrolyte under the conditions prevailing in the electrolysis. The most commonly used inert anodes for the corrosion protection of susceptible structures are graphite, silicon iron (14% silicon), titanium, or niobium plated with platinum or conductive polymers.

inert gases—former name for the noble gases. *See also* **noble gases**.

inertia—resistance offered by a body to a change of its state of rest or motion and being a fundamental property of matter.

inert material—loose term for a material that is either nonreactive or not very reactive chemically, e.g., noble metals, plastics, cement.

inert pigment—any nonreactive pigment or filler.

infinitesimal—vanishingly small but not zero.

inflammable—deprecated term meaning tendency to ignite easily and burn rapidly. *Same as* **flammable**, which is the preferred term.

infrared (IR)—that region of the electromagnetic spectrum lying beyond the visible red and having wavelengths from 750 nanometers to a few millimeters.

infrared spectroscopy (IR spectroscopy)—technique for chemical analysis and the determination of structure based on the principles that molecular vibrations occur in the infrared region and functional groups have characteristic absorption frequencies.

ingot—cast metal in a form intended for subsequent working or fabrication.

inhibitive pigment—pigment that prevents or retards the corrosion process.

inhibitive primer—metal primer paints that employ inorganic (rarely organic) inhibitive pigments to suppress corrosion reactions either directly (by their own solubility) or indirectly via the solubility of the products of their reactions with the vehicle or the products of the degeneration of the vehicle.

inhibitor—substance used to suppress a chemical reaction.

inhibitor, corrosion—*see* **corrosion inhibitor**.

inorganic—being composed of matter other than hydrocarbons and their derivatives; or matter that is not of plant or animal origin. *Contrast with* **organic**.

inorganic coatings—essentially cold-applied ceramic coatings, but not as dense or glossy as a fused ceramic coating. Also coatings employing inorganic binders or vehicles, such as silicates or phosphates, and that are usually pigmented with metallic zinc.

inorganic zinc-rich paint—coating containing a zinc powder pigment in an inorganic vehicle.

in situ—Latin term meaning in its natural or original position.

insolation—solar radiation received at the earth's surface per unit area.

insulated—describes being separated from other conducting surfaces by a dielectric substance (including air space) and offering a high resistance to the passage of current.

insulating oil—oxidation-resistant oil refined to give long service as a dielectric and coolant for transformers and other electrical equipment.

insulation—material used to prevent or retard passage of heat, electricity, or sound into or out of a body or region.

insulation resistance, electrical—ratio of the applied voltage to the total current between two electrodes in contact with a specific insulator.

insulator—substance that is a poor conductor of heat, electricity, or sound.

intentiostatic—*see* **galvanostatic**.

intercooling—improvement in cooling efficiency brought about by cooling between stages of compression.

intercrystalline corrosion—*see* **intergranular corrosion**.

intercrystalline cracking—*see* **caustic embrittlement**.

interdendritic—located within the branches of a dendrite or between the boundaries of two or more dendrites.

interdendritic corrosion—corrosive attack that progresses preferentially along interdendritic paths.

interface—common boundary between two materials.

interface corrosion inhibition—corrosion retardation mechanism that presumes a strong interaction between the corroding substrate and the corrosion inhibitor, which is potential-dependently adsorbed acting either as a positive or negative electrocatalyst on the corrosion reaction or where the inhibitor itself undergoes an electrochemical redox process. *Distinguished from* **interphase corrosion inhibition**.

interfacial tension—contractile force in a liquid-liquid phase interface.

interference color—color appearing in thin films illuminated with composite light and that depends on the angle of incidence, thickness of the film, and its refractive index.

intergranular—between crystals or grains.

intergranular corrosion—corrosion occurring preferentially at grain boundaries.

intergranular stress-corrosion cracking (IGSCC)—stress-corrosion cracking occurring along grain boundaries.

interionic attraction—electrostatic attraction between ions of unlike charge (sign).

interionic repulsion—electrostatic repulsion between ions of like charge (sign).

intermediates—organic compounds considered as chemical stepping stones between the parent substance and final product.

intermetallic compound—compound consisting of two or more metallic elements present in definite proportions in an alloy.

intermolecular forces—attractive forces among molecules that are responsible for holding them in a particular physical state. Weak forces occurring between molecules. *See also* **van der Waals force** and **hydrogen bond**.

internal corrosion—corrosion occurring inside a pipe because of the physical, chemical, or biological interactions between the pipe and water it transports, as opposed to forces acting outside the pipe, e.g., soil, weather, stress conditions.

internal oxidation—formation of particles of corrosion product beneath a surface. *Also known as* **selective corrosion** or **subsurface corrosion**.

internal phase—in an emulsion, the discontinuous phase. *See also* **discontinuous phase**. *Contrast with* **continuous phase**.

internal standard—material present in or added to samples in known amount to serve as a reference measurement.

International System of Units (SI)—principal system of measurement units used by scientists throughout the world.

International Union of Pure and Applied Chemistry Nomenclature System (IUPAC System)—systematic set of rules used to assign names to chemical compounds.

interphase corrosion inhibition—involves a three-dimensional layer between the corroding substrate and the electrolyte. The inhibitor layer generally consists of a weakly sol-

interstitial

uble compound such as oxidic corrosion products, intermediates, or inhibitors. Inhibition depends on the mechanical, structural, and chemical properties of the layer.

interstitial—occurring between the grains or in the pores of rock or soil. Also small atoms in spaces between the lattice atoms. They generally cause some distortion of the lattice resulting in local stress.

intumescent—swelling or bubbling of a coating because of heating.

invert emulsion—water-in-oil emulsion.

in vitro—event or process occurring outside a living organism or in an unnatural environment.

in vivo—event or process occurring naturally or spontaneously within a living organism or natural environment.

in vivo corrosion—corrosion of orthopedic devices and other surgical implants due to the action of body fluids.

iodine number—amount of iodine absorbed by an oil, fat, or wax giving a measure of the unsaturated linkages present.

ion—atom or radical in solution carrying an integral electric charge, either positive (cation) or negative (anion).

ion erosion—deterioration of a material caused by ion impact.

ion exchange—process by which certain undesired ions of given electric charge are absorbed from solution within an ion-permeable absorbent and being replaced in the solution by desirable ions of similar charge from the absorbent.

ion-exchange capacity—number of milliequivalents of exchangeable ions per (a) milliliter of backwashed and settled bed of ion-exchange material in its standard form (volume basis), (b) dry gram of exchange material in its standard form (weight basis).

ion-exchange membrane—ion-exchange material used as a barrier between two fluids.

ion-exchange resin—synthetic, organic ion-exchange material; usually containing carboxylic, phenolic, sulfonic, or substitute amino groups.

ionic bond—*see* **chemical bond**.

ionic crystal—crystalline compound consisting, effectively, of ions bound together by their electrostatic attractions, e.g., the alkali halides.

ionic migration—displacement of ions in solution under the effect of an electrical field.

ionic mobility—characteristic speed of a given ion in solution under an electric field of 1 V/cm. It depends on its dimension, charge, and the viscosity of the electrolyte.

ionics—electrochemical study of the interaction of ions under coulombic forces, of electrolytes, the displacement of ions under the influence of an electric field, and electrolytic conductivity. *See also* **electrodics**.

ionic strength—expression of the effect of the charge of the ions in a solution, equal to the sum of the molality of each type of ion present multiplied by the square of its charge.

ion implantation—process of modifying the physical or chemical properties of the near surface of a solid by embedding appropriate atoms into it from a beam of ionized particles.

ionization—process of producing ions; process by which a charged particle, either positive or negative, is created from a parent atom or molecule.

ionization potential—work (expressed in electron volts) required to remove a given electron from its atomic orbit and place it at rest at an infinite distance.

ionizing radiation—radiation that produces ions as it traverses matter; e.g., alpha particles.

ion-microprobe analysis—technique for analyzing the surface composition of solids by irradiating it with a beam of high-energy ions. Freed ions are analyzed in a mass spectrometer.

ionogens—substances that, although in the pure state are nonelectrolytic neutral molecules, can react with certain solvents to form products that rearrange to ion pairs which then dissociate to give conducting solutions, e.g., acetic acid and acetic acid in water.

ionomer—polymers in which ionized carboxyl groups form ionic cross links in the intermolecular structure.

ionophores—substances that exist only as ionic lattices in the pure crystalline form, and that when dissolved in an appropriate solvent give conductances that change according to some fractional power of the concentration, e.g., sodium chloride.

ion plating—generic term applied to atomistic film deposition processes in which the substrate surface and/or depositing film is subjected to a flux of high-energy particles, usually gas ions, sufficient to cause changes in the interfacial region or film properties. A physical vapor deposition coating process (PVD). Also called ion vapor deposition.

ion scattering spectroscopy (ISS)—elemental analytical technique similar to secondary ion mass spectroscopy (SIMS) in using an excited beam ion source and where the sample surface atoms scatter the incident ions, which are energy analyzed and compared with that of known elements for identification.

ion size—sum of the ionic radii of the oppositely charged ions in contact. Also called the distance of closest approach of the ions. Generally, the ion size is greater than the sum of the crystal radii.

ion vapor deposition—*see* **ion plating**.

IPCT—abbreviation for industrial process cooling tower.

ipy—abbreviation for inches per year.

IR—abbreviation for infrared radiation. *See also* **infrared**.

IR drop—voltage that results from the flow of current through a conductor.

iron—metallic element having three crystal forms: alpha-iron, which is stable below 906 °C (1663 °F) with a body-centered-cubic structure; gamma-iron, which is stable between 906 and 1403 °C (1663 and 2557 °F) with a nonmagnetic face-centered-cubic structure; and delta-iron, which is the body-centered-cubic form existing above 1403 °C (2557 °F). Commercially pure irons are obtainable as ingot iron.

iron fouling—fouling of tubes with iron oxide corrosion product.

iron phosphate coating—conversion coating formed on iron or steel surfaces by chemical reaction with a phosphoric acid solution.

iron rot—deterioration of wood from contact with iron.

irradiance—radiation incident on a surface expressed in watts per square meter. It is the total of the incident radiation at all wavelengths, whereas illuminance refers only to visible radiation. *See also* **spectral irradiance** and **spectral energy distribution**.

ISO—abbreviation for the International Organization for Standardization (Geneva).

ISO 7441 test—ISO atmospheric corrosion test for galvanic couples, where panels of the two metals are bolted together, atmospherically exposed, and the anodic metal later examined for degradation either visually, by weight loss, by tensile strength loss, or other.

isocorrosion diagram—graph or chart showing constant corrosion behavior with changing solution or other environment composition and temperature.

isoelectric point—pH value at which a substance or system is electrically neutral.

isomerism—chemical compounds (isomers) that have the same molecular formula but different molecular structures or different arrangements of atoms in space.

isotonic—solutions having the same osmotic pressure.

isotopes—atoms having the same atomic numbers but different mass numbers.

isotropic—substance whose physical properties are independent of direction.

ISS—*abbreviation for* **ion scattering spectroscopy**.

IUC—abbreviation for the International Union of Chemistry.

IUPAC—abbreviation for the International Union of Pure and Applied Chemistry. Sets forth systematic rules used to assign names to chemical compounds.

IVD—abbreviation for ion vapor deposition.

J K

japan—varnish yielding a hard, glossy, dark-colored film; usually dried by baking at relatively high temperatures.

jeweller's rouge—red, powdered hematite used as a mild abrasive for cleaning and polishing. *See also* **hematite**.

Jolly balance—spring balance for determining the specific gravity of a solid by weighing it alternatively in air and water.

joule—SI unit of work and energy equal to the work done when the point of application of a force of one newton moves in the direction of the force a distance of one meter. Symbol is J.

kaolin—soft white clay composed chiefly of the mineral kaolinite (essentially $Al_2O_3 \cdot 2SiO_2 \cdot 2H_2O$) used as a filler in rubber and paint. Also known as china clay.

Karl Fischer reagent—reagent used to conduct the Karl Fischer test. It contains iodine, sulfur dioxide, pyridine, and ethylene glycol monomethyl ether.

Karl Fischer test—laboratory procedure used to quantitatively measure the water or moisture content of organic liquids, solid resins, pigments, or coatings. It is based essentially on the reduction of iodine by sulfur dioxide in the presence of water when pyridine and an alcohol are also present. *See also* **Karl Fischer reagent**.

kelvin—symbol K. The SI (abbreviation for the International System of Units) unit of thermodynamic temperature equal to the fraction $1/273.16$ of the thermodynamic temperature of the triple point of water. The magnitude of the kelvin is equal to that of the degree Celsius.

kerogen—*see* **oil shale**.

kerosene—low-viscosity oil distilled from petroleum or shale oil. Commonly used for fuel, as a paint thinner, and as an alcohol denaturant.

Kesternich test—cabinet corrosion test utilizing a sulfur dioxide atmosphere to simulate a severe industrial environment.

ketones—organic compounds containing a carbonyl group bonded to two hydrocarbon groups (*RCOR'*).

keying—mechanical interlocking component between a phosphate conversion coating and its substrate metal.

kilo—prefix that is placed in front of units to increase their magnitude 1000 times.

kinetic energy—energy possessed by moving bodies.

kinetic friction—friction developed between two bodies in motion.

Kjeldahl nitrogen determination—analytical method for determination of nitrogen in an organic compound by its digestion and decomposition with sulfuric acid causing reduction to ammonium salts from which ammonia is liberated by a nonvolatile alkali and then distilled into a standard acid.

KLA—*abbreviation for* **knife-line attack**.

knife-line attack (KLA)—intergranular corrosion of an alloy, usually stabilized stainless steel, along a line adjoining or in contact with a weld after heating into the sensitization temperature range.

kraft process (kraft pulping)—alkaline wood-pulping process in which sodium sulfate is used as the caustic soda pulp-digestion liquor.

kraft pulp—pulp cooked by the alkaline liquor consisting essentially of a mixture of caustic soda and sodium sulfide.

Krebs units (KU)—units of Stormer viscosity applied to paints.

KTA panel—trademark for a paint test panel with unique configuration and markings.

L

labile—certain atoms or groups in a chemical compound that can easily be replaced by other atoms or groups.

lac—resinous substance secreted by a scale insect and used to make shellac.

lacquer—coating composition that is based on synthetic thermoplastic film-forming material dissolved in an organic solvent that dries primarily by solvent evaporation. Typical lacquers include those based on nitrocellulose, other cellulose derivatives, vinyl resins, acrylic resins, etc. Also a deposit resulting from the oxidation and polymerization of fuels and lubricants when exposed to high temperatures. Similar to but harder than varnish.

laitance—milky white deposit on new concrete; efflorescence.

lake—special type of pigment consisting essentially of an organic soluble coloring matter (dyestuff) combined with an inorganic base or carrier.

Lambert's law of absorption—each layer of equal thickness absorbs an equal fraction of the light that traverses it.

lamellar corrosion—*see* **exfoliation corrosion**.

lamellar solid—solid substance in which the crystal structure has distinct layers, e.g., mica, graphite.

laminar flow—situation where nonturbulent fluid moves in parallel lamina or layers. *See also* **streamline flow**.

laminar scale—rust formation in heavy layers.

laminate—sheets of material united by a bonding material usually with pressure and heat.

lamp black—finely divided form of carbon made by burning organic compounds in insufficient oxygen.

Langelier saturation index (LSI)—index calculated from total dissolved solids, calcium concentration, total alkalinity, pH, and solution temperature that shows the tendency of a water solution to precipitate or dissolve calcium carbonate. The Langelier saturation index is given by the relation: $L_{SI} = pH_{meas} - pH_s$, where L_{SI} is the Langelier saturation index, pH_{meas} is the measured pH, and pH_s is the saturation pH. The Langelier index is used to indicate water aggressiveness and the possibility for scaling. It is not, however, a corrosion index, and not always a predictor of the formation of protective scaling because the calcium salt may precipitate out as powder and not cover the metal. *See also* **calcium carbonate saturation index** and **corrosion index**.

lanolin—emulsion of purified wool fat in water.

lanthanides—fifteen elements of the periodic table that range in proton number from lanthanum (57) to lutetium (71), inclusive. *Also called* **rare earths**.

lapping—abrasive removal of material using abrasive particles in a liquid slurry.

lard oil—animal oil made from chilled lard or from the fat of swine.

LAS—abbreviation for linear alkylsulfonates, which are used as biodegradable detergents.

latent solvent—liquid unable to dissolve a binder, but that increases the tolerance of paint for a diluent. *See also* **active solvent**.

latex—stable dispersion of polymeric substances in an essentially aqueous medium. Natural and synthetic latices (latexes) are available. Natural latex is a milky fluid of mixed composition found in some herbaceous plants and trees.

latex-modified Portland cement concrete—concrete that includes a polymer emulsion to produce a dense-hardened concrete that resists the movement of moisture and chloride ions.

latex paint—paint containing colloidal binder particles formed by emulsion polymerization as the principal constituent of the binder.

lattice—regular arrangement of atoms, ions, or molecules in a crystalline solid.

lattice compounds—compounds formed between definite stoichiometric amounts of two molecular species and that owe their stability to packing in the crystal lattice, not to valence forces.

law of conservation of mass—mass cannot be created or destroyed in normal chemical changes.

law of constant composition—mass ratios of elements within a compound are fixed.

law of mass action—rate at which a chemical reaction takes place at a given temperature is proportional to the product of the active masses of the reactants. The active masses of the reactants are their molar concentrations.

layer corrosion—corrosion attack of a metal surface that is localized in thin parallel layers oriented in the direction of metal processing and that leads to unattached metal layers being released like pages in a book.

LC—abbreviation for lethal concentration. LC_{50} stands for the concentration that is lethal to 50% of the animals tested. *Also abbreviation for* **liquid chromatography**.

LD—abbreviation for lethal dose. LD_{50} stands for the dose that is lethal to 50% of the animals tested.

LDPE—*abbreviation for* **low-density polyethylene**.

leaching—removal of more soluble components from a mixture with an insoluble component by percolating or by a moving solvent, usually water.

lead phosphite, dibasic—$2PbO \cdot PbHPO_3 \cdot \frac{1}{2}H_2O$. A white corrosion inhibitive paint pigment also used to retard bleed-through of cedar and redwood stains in latex paint.

lead storage battery—consists of a lead electrode and a lead oxide electrode immersed in sulfuric acid. The cell is Pb; $PbSO_4$, H_2SO_4 (20%); PbO_2, Pb.

leafing—orientation of pigment flakes in horizontal planes.

Le Chatelier's principle—if a system is in equilibrium, any change imposed on the system tends to shift the equilibrium to nullify the effect of the applied change.

Legionnaire's Disease—name given to diseases that may result from the Legionella bacteria invading humans through breathed vapor from sources where it has multiplied greatly such as in cooling towers, hot-water heating systems and other places where water stands or is recirculated. The bacteria invades and kills the lung's white blood cells that normally kill most other bacterial invaders and can cause respiratory diseases, diarrhea, kidney and liver problems, deliriousness, and hallucinations. Sources where the bacteria might be accumulating are generally decontaminated with chlorine treatments.

leonardite—naturally occurring oxidized lignite.

ligand—any molecule, ion, radical, or group in a complex (polyatomic group) bound to the central atom, e.g., the ammonia molecules in $[Co(NH_3)_6]^{3+}$. Ligands are also complexing agents. *See also* **chelate** *and* **complexing agent**.

light blast cleaning—*see* **white blasting**.

lightfastness—relative resistance of materials to change in color.

light water—in nuclear power terminology, it is the term for ordinary water. *Contrast with* **heavy water**.

light-water reactor (LWR)—type of fission reaction reactor that may be either of two principal designs: pressurized-water reactor (PWR) or boiling-water reactor (BWR). *See also* **boiling-water reactor** and **nuclear pressurized-water reactor**.

lignin—major (about 25%) noncellulose constituent of wood and all vascular plants. A complex aromatic organic polymer that is deposited within the cellulose of plant cell walls during secondary thickening.

lignite—form of coal that is soft, brown, and has a high moisture content.

lignosulfonates—organic drilling fluid additives derived from by-products of sulfite paper manufacturing processes from coniferous woods. Most common lignosulfonate salts are the ferrochrome, chrome, calcium, and sodium.

lime—calcium oxide (CaO), or a mixture of calcium oxide and magnesium oxide (MgO). *Also called* **quicklime**. Also the name given to the commercial form of calcium hydroxide ($Ca(OH)_2$).

lime-base grease—grease prepared from a lubricating oil and a lime soap.

lime softening—process for reducing the amount of calcium bicarbonate in water through the precipitation of calcium carbonate. It also reduces magnesium alkalinity by precipitating magnesium hydroxide. Cold lime softening is carried out at ambient temperature, warm lime softening on preheated water at 50 to 66 °C (120 to 150 °F), and hot lime softening at 100 °C (212 °F) and higher.

limestone—sedimentary carbonate rock composed chiefly of calcite ($CaCO_3$).

limiting current density—maximum current density that can be used to obtain a desired electrode reaction without undue interference, such as from polarization.

limiting diffusion current density—at very high hydrogen ion reduction rates, the region adjacent to the electrode surface becomes depleted of hydrogen ions. Upon further increase in reduction, a limiting rate is reached as determined by the diffusion rate of hydrogen ions to the electrode surface. This rate is the limiting diffusion current density and it represents the maximum rate of reduction possible for a given system.

limit of detection (LOD)—concentration of pollutants as regulated at the limit of their detection.

limit of quantitation (LOQ)—defined by the ACS as 10 standard deviations away from the average signal calculated from replicate analyses of the blank.

linear polarization measurements—instantaneous corrosion rate determination using test probes inserted into electrically conductive process streams or in containers of electrically conductive process fluids.

linseed oil—pale yellow drying oil pressed from seeds of the flax plant. It contains a mixture of glycerides of fatty acids, including linoleic acid and linolenic acid. It is the most widely used oil in the paint industry and one that is characterized by its relatively short drying time.

lipophilic—having an affinity for oil; *opposite of* **hydrophilic**.

liquid—any substance that has a definite volume but no definite form except as defined by its container and that flows readily or changes shape in response to the smallest force. Matter in a fluid state but relatively incompressible. *Contrasts with* **solid**.

liquid-applied linings—polymeric coatings that may be applied by spraying or troweling.

liquid chromatography (LC)—chromatographic technique used for the separation and determination of materials in a liquid sample.

liquid-impingement erosion—loss of material from a solid surface due to continued exposure to impacts by liquid drops or streams.

liquid junction potential—difference of potential that appears in the region between two ionic solutions of different concentrations.

liquid-metal corrosion—corrosion of solids exposed to liquid-metal environments. Relevant corrosion phenomena include: dissolution, impurity and interstitial reactions, alloying, and compound reduction.

liquid-metal cracking (LMC)—liquid-metal penetration along the grain boundaries of some metal systems resulting in cracking. Usually does not involve corrosion. *Also known as* **liquid-metal embrittlement**.

liquid-metal embrittlement (LME)—form of environmental cracking resulting in catastrophic brittle failure of a normally ductile metal brought on from contact with a liquid metal and subsequently being stressed in tension. Also known as liquid-metal assisted cracking.

liquid-penetrant inspection—nondestructive means for detecting surface discontinuities by applying a liquid penetrant, removing the excess, then applying a "developer" to act as a blotter to draw out a portion of the penetrant seeped into the surface openings. Residual penetrant, usually a visible dye, indicates any discontinuity.

liter (L)—non-SI unit of volume, equivalent to the SI unit $1 \text{ d} \cdot \text{m}^3$. Volume of a kilogram of water at 4 °C (39 °F) and equal to 1.06 quarts.

litharge—inorganic compound PbO.

lithium-base grease—grease prepared from a lubricating oil and a lithium soap.

lithosphere—solid portion of the Earth.

lithium bromide—LiBr; very hygroscopic compound whose primary use is in absorption-refrigeration air-conditioning systems in which water is the refrigerant. LiBr is used to absorb water vapor.

litmus—water-soluble dye that turns red under acid conditions and blue under alkaline conditions with the color change occurring over the pH range 4.5 to 8.3. Litmus paper is absorbent paper soaked in litmus solution. *See also* **indicator**.

live oil—crude oil containing gas and that has not been stabilized or weathered and constitutes a potential fire hazard.

LMC—*abbreviation for* **liquid-metal cracking**.

LME—*abbreviation for* **liquid-metal embrittlement**.

LNG—abbreviation for liquefied natural gas.

loam—mixture of sand, silt, clay, or a combination of these with organic matter (humus).

LOCA—abbreviation for loss of coolant accident.

local action—corrosion due to the action of local cells and caused by nonuniformities between two adjacent areas at a metal surface exposed to an electrolyte.

local cell—galvanic cell resulting from local action.

local corrosion cell—electrochemical cell on a metal surface developed because of a potential difference between adjacent areas on that surface.

localized corrosion—corrosion at discrete sites, e.g., crevice corrosion, pitting, intergranular attack, corrosion fatigue, and stress-corrosion cracking.

LOD—*abbreviation for* **limit of detection**.

long-line current—current flowing through the earth from an anodic to a cathodic area that returns along an underground metallic structure.

long oil—resin having a large quantity of oil, i.e., more than 25 gallons per 100 pounds of resin.

long terne coated steel—lead-tin alloy (usually 3 to 8% tin) coated steels offering corrosion protection for automotive gas tanks, fuel lines, and brake lines. The terne is cathodic to the steel substrate.

loose flake—thin, easily removed mill scale.

LOQ—*abbreviation for* **limit of quantitation**.

low alloy steel—carbon steel alloyed, singly or in combinations, with chromium, nickel, copper, molybdenum, phosphorus, and vanadium in the range of a few percent or less.

low-boiling-point engine coolant—mixture of alcohol engine coolant concentrate and water. *See also* **high-boiling-point engine coolant**.

low-density polyethylene plastics (LDPE)—branched polyethylene plastics having a density of 0.910 to 0.925 g/cm^3.

low-pressure spraying—conventional air spraying.

low-solids mud—drilling mud in which additives, e.g., carboxymethyl cellulose, have been partially or wholly substituted for commercial or natural clays.

LPG (LP gas)—abbreviation for liquefied petroleum gas composed principally of propane and butane and stored as a liquid under pressure.

LSI—*abbreviation for* **Langelier saturation index**.

lubricant—any substance interposed between two surfaces in relative motion for the purpose of reducing the friction and/or wear between them.

lubricating grease—lubricating oil to which material has been added to thicken it to a semifluid or solid state, and which material may also contain additives to enhance specific properties such as its oxidation stability, rust-preventive ability, and extreme-pressure capability.

luggin probe (Luggin-Haber capillary)—small tube or capillary filled with electrolyte and terminating close to the metal surface under study. Used to provide an ionically conducting path, without diffusion, between an electrode under study and a reference electrode.

lumber—trees or wood sawed into boards, planks, or structural members of standard or specified length.

lustrous—having a high degree of specular reflectivity.

LWR—*abbreviation for* **light-water reactor**.

lye—see **caustic soda**.

lyophilic—having an affinity for a solvent, i.e., solvent-loving. If the solvent is water, the term hydrophilic is used.

lyophobic—lacking any affinity for a solvent, i.e., solvent-hating. If the solvent is water, the term hydrophobic is used.

M

MAC—abbreviation for maximum allowable concentration. The term has generally been replaced by the term threshold limit value. *See also* **threshold limit value (TLV)**.

macrograph—graphic reproduction of an object either unreduced in size or magnified no more than about ten diameters.

macromolecule—molecule with a very high molecular mass (usually an aggregation of hundreds or thousands of atoms). Natural and many synthetic polymers have macromolecules.

macroscopic—visible at magnifications to about 25×.

macrostructure—structure as revealed by macroscopic examination.

magnesia—magnesium oxide (MgO).

magnesium hardness—magnesium concentration of a water, but expressed as calcium carbonate ($CaCO_3$), concentration. *See also* **total hardness**.

magnet—body that produces a magnetic field external to itself.

magnetic field—field of force that exists around a magnetic body or a current-carrying conductor. This field of force may cause a magnetic dipole within it to experience a torque and a moving charge may experience a force. *See also* **electromagnetism**.

magnetic particle inspection—nondestructive procedure for detecting surface discontinuities in ferromagnetic materials by establishing a magnetic field in the material object, applying magnetic particles to the surface, examining the surface for accumulation of the particles, and evaluating the object for serviceability.

magnetism—group of phenomena associated with magnetic fields. Whenever an electric current flows, a magnetic field is produced.

magnetite—spinel-type naturally occurring black magnetic iron oxide. It is a mixed iron(II)-iron(III) oxide (Fe_3O_4). Also formed as a component of scale on steel and that is harder to dissolve by pickling acids than is the wustite (FeO) component of scale. *See also* **scale** and **wustite**.

magnetostrictive cavitation test device—vibratory cavitation test device driven by a magnetostrictive transducer.

magnification—measure of the extent to which an optical system enlarges or reduces an image.

maintenance painting—painting to maintain plants, offices, stores, and other commercial structures, hospitals, schools, public and private buildings, and such areas as public utilities, roadways, etc., with the primary function of protection. Also any repair painting after the initial paint job.

makeup water—original water that enters the industrial process and may include municipal water, well water, surface water, or condensate water. It is usually further purified for specific use in boilers. Also the water supplied to a cooling system to compensate for water loss by evaporation and any leakage.

maleic anhydride copolymers—typically, maleic methylvinyl ether and styrene maleic copolymers. Used in cooling waters for scale and fouling control.

malleable—property of substances that enables them to be hammered and shaped into different forms; a distinctive property of metals.

malleable iron—white iron that has been heat treated to convert its metallic matrix to contain nodules of temper carbon. Shows good ductility.

mandrel test—physical bending test for evaluation of coating adhesion and flexibility.

manganese bronze—similar to Muntz metal but also containing about 1% each of tin, iron, and lead. It is primarily used for ship propellers. *See also* **Muntz metal**.

man-made air pollution—air pollution resulting directly or indirectly from human activities.

manometer—instrument used to measure pressure differences by the difference in height of two liquid columns. It consists of a static liquid column that is used to balance the pressures applied to the two ends of the column.

manufactured sand—fine aggregate produced by crushing rock, gravel, slag, or concrete.

maraging—precipitation-hardening treatment applied to a special group of iron-base alloys to precipitate one or more intermetallic compounds in a matrix of essentially carbon-free martensite.

marble—carbonate rock that has acquired a distinctive crystalline texture by recrystallization during metamorphism and is composed principally of calcite ($CaCO_3$).

marine—of or pertaining to the sea.

marine borers—mollusks and crustaceans that attack submerged wood in salt and brackish water.

marine coatings—paints and varnishes specifically formulated to withstand water immersion or exposure to a marine atmosphere.

marine environment—atmospheric exposure that is frequently wetted by salt mist, but which is not in direct contact with salt spray or splashing waves.

marine oils—liquid organic substances of a high degree of unsaturation obtained from the fatty tissue of fish, whales, and other marine animals. Used as boundary lubricants.

marsh gas—methane (CH_4) formed by rotting vegetation in marshes.

martensite—generic name for a metastable phase resulting from the diffusionless athermal decomposition of austenite below a certain temperature known as the M temperature (martensite start temperature) and characterized by an acicular pattern in its microstructure. *See also* **austenite**.

martensitic stainless steels—stainless steels containing about 12 to 18% chromium, 0.08 to 1.10% carbon, and small amounts of one or more other elements such as nickel, niobium, molybdenum, selenium, silicon, and sulfur. These steels are characterized by their martensitic structure, which is obtained by rapid quenching, and by their ability to be hardened by the same heat treatments used to harden carbon and alloy steels. They are usually less corrosion resistant than the austenitic and ferritic stainless steels, and like the ferritic stainless steels they are in the "400" series of stainless steels, with the most common being type 410.

masonry—stonework or brickwork. Construction, usually set in mortar, of natural building stone or manufactured units such as brick, concrete block, adobe, glass block, tile, manufactured stone, or gypsum block.

masonry cement—hydraulic cement for mortars used in masonry construction and containing one or more of various Portland cements and hydraulic limes.

masonry paints—alkali-resistant coating, usually a latex paint, used for masonry substrates.

mass—quantity of matter in a body. A measure of a body's inertia, i.e., its resistance to acceleration.

mass number—total number of nucleons in the nucleus of an atomic species.

mass spectroscopy—technique used to determine relative atomic masses and the abundance of isotopes, and that is also used for chemical analysis and the study of ion reactions.

mastic—heavy-bodied, high-build coating or adhesive composition.

mat—randomly distributed felt of glass fibers used in reinforced plastics lay-up molding.

material—relating to, derived from, or consisting of matter.

matrix—continuous solid phase in which particles are embedded.

matrix metal—continuous phase of a polyphase alloy or mechanical mixture.

matte—being without gloss or luster but having a smooth surface to the touch.

matter—anything that has mass and occupies space.

Mattsson's solution—aqueous solution of 0.05 g-atom/L of Cu^{2+} and 1 g-mol/L of NH_4^+ with a pH of 7.2 used in ASTM G 37 for the evaluation of stress-corrosion cracking of brasses.

McCauley's driving force index (DFI)—index based on $CaCO_3$ solubility used to predict the amount of $CaCO_3$ that will precipitate from a water.

MCL—abbreviation for maximum contaminant level.

mdd—abbreviation for milligrams per square decimeter per day.

MDL—abbreviation for method detection level.

mean—arithmetical average of a set of numbers.

mechanical bond—adherence of a coating to a base material mainly through mechanical interlocking with roughened surfaces.

mechanical draft cooling towers—cooling towers where air is introduced by either forced or induced draft. Designed to minimize water loss from windage and drift.

mechanical galvanizing—*see* **mechanical plating**.

mechanically assisted degradation—any type of degradation involving both corrosion and a wear or fatigue mechanism. Examples include: erosion, fretting, fretting fatigue, cavitation, impingement, and corrosion fatigue.

mechanical plating (mechanical coating)—process whereby hard, small spherical objects, such as glass shot, are tumbled against a metallic surface in the presence of finely divided metal powder, such as zinc dust, and appropriate chemicals for the purpose of covering the metallic surface with the powdered metal. Also referred to as peen plating, impact plating, and mechanical galvanizing.

mechanical properties—properties of a material that are associated with elastic and inelastic reaction when force is applied, or that involve the relationship between stress and strain. This definition is sometimes included in the term physical properties, but the term mechanical properties is preferred.

mechanical refrigeration—any of the processes for lowering the temperature of a substance below its surroundings by using a fluid that evaporates and condenses at suitable pressures. All mechanical refrigeration results in the simultaneous production, somewhere else, of a greater amount of heat. Widely used by the chemical process industry to remove the heat of chemical reactions, to liquefy process gases for gas separation by distillation and condensation, to purify products by preferential freeze-out of one component from a liquid mixture, and to air-condition plant areas.

mechanics—study of the interactions between matter and the forces acting on it.

mechanism—way in which a particular chemical reaction occurs and is usually described in terms of the steps involved.

median—after arranging a series of observations in their order of magnitude, the value that falls in the middle when the number of observations is odd, or the arithmetic mean of the two middle observations when the number of observations is even.

medium—surrounding environment in which something functions or exists.

medium density polyethylene plastic (MDPE)—branched polyethylene plastic having a density of 0.926 to 0.940 g/cm^3.

melamine resins—synthetic resins that are condensate products of formaldehyde and melamine and that require baking for cure.

melting point (mp)—temperature at which a solid changes into a liquid.

melt processible rubber—alloys of ethylene interpolymers and chlorinated polyolefins.

membrane—any thin sheet or layer. Also a barrier, usually thin, that permits the passage only of particles up to a certain size or of special nature.

meniscus—curved upper surface of a liquid column that is concave when the containing walls are wetted by the liquid and convex when not.

mer—smallest repeating structural unit in a polymer.

mercaptans—*see* **thiols**.

mercury corrosion—specific example of liquid-metal corrosion where the liquid metal is mercury. Mercury is particularly corrosive toward stressed hard brass and aluminum. *See also* **liquid-metal corrosion**.

mercury droplet wetting test—test for cleanliness of metal surfaces. Droplet of mercury placed upon a clean metal surface will spread while one that has grease or oxide present will cause the droplet to retain a roughly spherical shape.

mesh—square opening of a sieve. A measure of fineness of a woven material, screen, or sieve. For example, a 100 mesh sieve has 100 openings per linear inch.

metabolize—to convert food, such as organic matter, to cellular matter and gaseous byproducts by a biological process.

metal—class of chemical elements that are typically ductile, fusible, opaque, lustrous, crystalline, and that conduct electricity. However, not all metals have all these properties. Metals tend to form positive ions in contrast with nonmetals which, typically, form negative ions. All metals have oxides that are basic.

metal blister—bloating of metal sheet.

metal dusting—form of high-temperature corrosion that produces a dustlike corrosion product in the presence of carbonaceous gases.

metal fatigue—cumulative effect causing a metal to fail after repeated applications of stress, none of which exceeds its ultimate tensile strength.

metal ion concentration cell—*see* **concentration cell**.

metal ion deposition—process in which ions of a more noble metal are reduced on the surface of a more active metal with the resulting metallic deposit providing cathodic sites for further galvanic corrosion of the more active metal. *See also* **cementation**.

metallic bond—chemical bond of the type holding together the atoms in a solid metal or alloy. In metals and alloys, the atoms are considered to be ionized, with the positive ions occupying lattice positions. The valence electrons are able to move freely through the lattice. The bonding force is electrostatic attraction between the positive metal ions and the electrons. The free electrons account for the good electrical and thermal conductivities of metals.

metallic composites—metallic compositions consisting of a metallic matrix strengthened by metallic or nonmetallic fibers, filaments, or whiskers. Characterized by greater strength and stiffness than the matrix metal and often more resistant to crack growth.

metallic crystal—crystalline solid in which the atoms are held together by metallic bonds.

metallic driers—metallic or organometallic compounds used to accelerate drying of oil, paint, printing ink or varnish.

metallic glass—metallic alloy having an amorphous or glassy structure and characterized by high strength, ductility, modulus of elasticity, and corrosion resistance. *See also* **amorphous alloy**.

metallic paint—paint that gives a film with a metallic appearance, usually achieved by incorporating fine flakes of copper, bronze, or aluminum.

metallic pigments—metals, except for zinc dust, in the form of flakes or flat platelets used as pigments in coatings. They tend to reinforce the binder, reinforce adhesion, and may also be used for metallic sheen or color. In the case of leafing metallic pigments, they create a shingle effect preventing actinic rays from penetrating into the binder to cause premature degradation. Common leafing pigments include aluminum, stainless steel, lead, and copper.

metallic soaps—salts derived from metals and fatty (organic) acids.

metallizing—process of coating a surface with a layer of metal. Also refers to the application of an electrically conductive metallic layer to the surface of nonconductors and to the application of metallic coatings by nonelectrolytic procedures such as spraying of molten metal and deposition from the vapor phase.

metallography—microscopic study of the structure of metals and their alloys.

metalloids—any of a class of chemical elements intermediate in properties between metals and nonmetals, e.g., boron, silicon, germanium, arsenic, and tellurium. They are electrical semiconductors, and their oxides are amphoteric.

metallurgical bond—adherence of a coating to the base metal characterized by diffusion, alloying, or intermolecular or intergranular attraction at the interface.

metallurgically influenced corrosion—corrosion that is affected by alloy chemistry and heat treatment factors such as the relative stability of the components of an alloy, metallic phases, metalloid phases such as carbides, and local variations in composition. The most common form of metallurgically influenced corrosion is intergranular corrosion that occurs when corrosion is localized at grain boundaries.

metallurgy—science and technology concerned with the production of metals from their ores, purification of metals, manufacture of alloys, and the use and performance of metals in engineering practice.

metal primer—primer or the first coat of paint on a metal.

metalworking fluids—fluids used to provide lubrication, cooling, and rust protection when machining, grinding, cutting, drawing, or forming metals. They include straight mineral oils or oil mixtures containing additives to enhance performance, emulsifiable or soluble oils that are suspensions of mineral, paraffinic, or naphthenic oils in water formulated by blending the oil with emulsifying agents and other materials, and synthetic fluids that contain no oily phase and are true aqueous solutions that typically utilize wetting agents, extreme-pressure agents, lubricity additives, corrosion inhibitors, and filming agents in their formulation. The emulsified oils are also called semisynthetics and soluble oils, while the synthetic fluids are also sometimes called chemical fluids.

metathesis—*see* **double decomposition**.

meter (m)—SI unit of length; 1 m = 39.37 in.

method—form of standard covering a precise procedure or technique used for performing inspection that may be specified.

method detection limit—defined by the EPA as the minimum concentration of a substance that can be measured and reported with 99% confidence that the analyte concentration is greater than zero and is determined from replicate analysis of a sample in a given matrix containing the analyte.

methyl cellulose—any of several methyl ethers of cellulose.

methyl orange—water-soluble color indicator of pH. Color is yellow above pH 4.4 and salmon pink below pH 4.4.

metric system—decimal system of units on which SI units are based. *See also* **SI**.

mg/L—milligrams per liter.

mho—reciprocal ohm. *See also* **ohm**.

MIC—*abbreviation for* **microbiologically influenced corrosion**.

mica—any of a group of naturally occurring silicate minerals with a highly perfect basal cleavage (layered crystal structure). Chemically, an alkali aluminum silicate.

micaceous iron oxide—naturally occurring iron oxide resembling mica in its leaflike structure and sometimes used as an anticorrosive pigment.

micelle—electrically charged colloidal particle, usually organic in nature, composed of many aggregated small molecules.

microbiologically influenced corrosion (MIC)—corrosion caused by the presence and activities of microorganisms within biofilms developed on surfaces in contact with aqueous environments. The reactions are usually localized and can include sulfide production, acid production, ammonia production, metal deposition, and metal oxidation and reduction. *See also* **bacteriological corrosion** and **biological corrosion**.

microcrystalline wax—plastic, high-melting point petroleum wax obtained from petrolatum by solvent extraction or by other means. The waxes are characterized not only by their microcrystalline structure but also by their very high average molecular weight and a much higher viscosity than that of paraffin wax.

microemulsion—transparent solution of water and oil that is thermodynamically stable.

micrograph—graphic representation of an object as viewed by a microscope or equivalent optical instrument at magnifications greater than about ten diameters.

micrometer—one millionth of a meter; 0.001 mm.

micron—obsolete term for micrometer. Unit of length equal to one millionth part of a meter.

micronutrient—substance that, in minute amounts, is essential to life.

microorganism—organisms, called microbes, observable only through a microscope. Larger organisms are called macroorganisms.

microprobe analysis—irradiation of a sample surface with a high-energy electron beam and using the emitted X-ray radiation to make chemical analysis of the surface possible. Used on elements whose atomic numbers are above 10.

microstructure—structure of a suitably prepared specimen as revealed by a microscope.

microwave spectroscopy—technique for chemical analysis and determination of molecular structure (bond angles and dipole moments) and also relative atomic masses.

migration—that displacement of ions in solution under the effect of an electrical field and whose speed of migration is called mobility. Mobility varies with different ions and also depends on the viscosity of the electrolyte.

mil—one thousandth of an inch; 0.001 in.

mildew—superficial coating or discoloring of organic materials, such as paint, cloth, leather, and the like, caused by fungi or mold, especially formed under damp conditions of use or storage.

mildew resistance—ability of a coating to resist fungus growth that can cause discoloration and ultimate decomposition of a coating's binding medium.

mildewstat—chemical agent that inhibits the growth of mildew. Also called mildewcide.

mild steel (carbon steel)—inexact term generally intended to denote a low-carbon steel, such as SAE 1010 steel; a common structural steel. It is an iron alloy with up to 2% added carbon that occupies the interstitial sites in the lattice structure and improves the mechanical properties of the steel by forming an insoluble second phase known as carbides with the iron. *See also* **carbon steel**.

milk emulsion—an oil-in-water emulsion.

Miller number—an index of the relative abrasivity of slurries. A measure of slurry abrasivity as related to the instantaneous rate of mass loss of a standard metal wear block at a specific time on the accumulative abrasion-corrosion time curve. See ASTM G 75 for test method details.

milli—prefix placed in front of a unit to diminish its size by $1/1000$.

millimeter of mercury (mm Hg)—non-SI unit of pressure equal to $1/760$ of an atmosphere.

mill scale—heavy oxide layer on iron and steel formed during heat treatment or hot working; most often associated with formation of magnetite. Sometimes referred to as blue scale.

mill scale binder—gray oxide layer between mill scale and steel.

mils per year (mpy)—an expression for corrosion rate in terms of penetration or thinning of a metal. Calculated from the weight loss of a metal specimen during a corrosion test by the formula: $mpy = 534W/DAT$, where W is the weight loss in milligrams, D is the density of the metal in grams per cubic centimeter, A is the area of the specimen in square inches, and T is the exposure time in hours. Some metric penetration rates calculated from mpy are: 1 mpy = 0.0254 mm/yr = 2.90 nm/hr = 0.805 pm/sec, where mm is 10^{-3} meter, nm is 10^{-9} meter, and pm is 10^{-12} meter.

MIL-STD—abbreviation for U.S. military standard.

mineral—any inorganic material having a definite chemical composition and structure that is found in a natural state. Incorrectly, but often applied to fossilized organic matter, such as coal, petroleum, and natural gas. These are not minerals, being complex mixtures without definite chemical formulas.

mineral acid—common strong acid such as hydrochloric (HCl), sulfuric (H_2SO_4), and nitric (HNO_3) so-called because they are derived from inorganic materials.

mineral fiber—fibers manufactured from rock, slag, or glass and with or without binders.

mineral jelly—*see* **petrolatum wax**.

mineral oil—any liquid product of petroleum within the viscosity range of products called oils and consisting of high-molecular-weight hydrocarbons. Classified as either naphthenic, with composition consisting essentially of saturated, ring-type structures, or paraffinic, with a composition consisting essentially of straight, or branched-chain saturated hydrocarbons.

mineral spirits—general term for refined petroleum distillate having a low aromatic hydrocarbon content; used as a thinner and solvent for paints.

mineral water—naturally occurring or prepared water that contains dissolved minerals or gases, often used therapeutically.

mineral wool—any inorganic fibrous material produced by steam blasting and cooling molten silicate or similar substance. Used as an insulator and filtering medium.

minimum determinability—lowest value that can be determined within the stated precision of a test method expressed quantitatively in the same dimension that is used for reporting results of the test.

misch metal—alloy of cerium (50%), lanthanum (25%), neodymium (18%), praseodymium (5%), and other rare earths; used to improve malleability of iron, make copper alloys harder, aluminum alloys stronger, reduce creep of magnesium alloys, and reduce oxidation of nickel alloys.

miscible—capable of mixing or blending uniformly.

mist—liquid, usually water, in the form of particles suspended in the atmosphere at or near the surface of the earth; distinguished from fog by being more transparent or having particles perceptibly moving downward. Also an aerosol with liquid as the dispersed colloid.

mist coat—very thin sprayed coat of the topcoat prior to application of a full heavy coat. Usually practiced when topcoating zinc-rich primer coats.

mixed bed—physical mixture of anion-exchange and cation-exchange materials.

mixed film lubrication—lubrication governed by a combination of boundary and hydrodynamic or elastohydrodynamic effects.

mixed potential—potential of a specimen (or specimens in a galvanic couple) when two or more electrochemical reactions are occurring.

mixed potential diagram—diagram allowing for discussion of electrode processes for currents higher than the corrosion current and providing for a rapid technique for determining the corrosion rate of metals and alloys in specific liquid environments. It is a plot of potential versus log of the current density.

mixed potential theory—states that during the corrosion of an electrically isolated metal sample, the total rate of oxidation must equal the total rate of reduction.

mixing—thorough intermingling of two or more materials.

mixture—system of two or more distinct chemical substances; a physical combination.

mm/a—abbreviation for millimeters per year.

mm Hg—unit of pressure equal to that exerted under standard gravity by a height of one millimeter of mercury.

modified salt spray (fog) test—see ASTM G 85 for details.

modifier—chemically inert ingredient added to a resin formulation that changes the properties of the resin.

modulus of elasticity—ratio of stress to strain of a material in the elastic region. A measure of its stiffness or rigidity.

Mohs scale—hardness scale in which a series of ten minerals are arranged in order; each mineral listed being scratched by and therefore softer than those numerically listed above it. The minerals are: (1) talc, (2) gypsum, (3) calcite, (4) fluorite, (5) apatite, (6) orthoclase, (7) quartz, (8) topaz, (9) corundum, and (10) diamond.

moiety—indefinite portion of a sample.

moist—slightly wet or damp; humid; filled with moisture.

moist sulfur dioxide test—see ASTM G 87 for details. This is a U.S. equivalent of the Kesternich test.

moisture—dampness. Diffuse wetness that can be felt as vapor in the atmosphere, or as condensed liquid on surfaces of objects.

moisture, atmospheric—ambient humidity that may be absorbed by hygroscopic material.

moisture resistance—ability of a material to resist absorbing moisture, either from the atmosphere or when immersed in water.

moisture vapor transfer rate (MVTR)—rate at which moisture vapor will transfer through a protective coating when there is a difference in moisture vapor pressure on one side of the coating compared with the other side.

moisture vapor transmission—rate of movement of moisture vapor through a membrane.

molal solution—solute concentration of a solution expressed as moles of solute divided by 1000 grams of solvent.

molar conductivity—conductivity of that volume of an electrolyte that contains one mole of solution between electrodes placed one meter apart.

molar solution—solution containing one mole (one gram-molecular weight) of solute in one liter of the solution.

molds—indefinite term for some fungi classified as saprophytes (i.e., obtaining their food from nonliving organic material), which play an important role in food spoilage and the biodegradation of various materials such as leather, paper, and wood products. The most critical factor in mold growth is the availability of sufficient moisture.

mole—one mole is the mass numerically equal, in grams, to the relative molecular mass of a substance. Quantity of a substance, expressed in specific mass units, that is equal to the molecular weight of the substance. It is the amount of substance of a system containing as many elementary units (6.02×10^{23}) as there are atoms of carbon in 0.012 kilograms of the pure nuclide ^{12}C.

molecular beam—beam of atoms, ions, or molecules at low pressure in which all the particles are travelling in the same direction with few collisions between them. Used in studies of surfaces, chemical reactions, and in spectroscopy.

molecular compound—chemical compound formed by union of two or more already saturated molecules, e.g., double salts, salts with water of crystallization, and metal ammonium derivatives.

molecular formula—chemical formula that indicates the type and exact number of atoms within a molecule.

molecularly dehydrated phosphates—phosphate compounds formed by molecular dehydration of their water constitution, i.e., the water that forms an integral part of the molecule in contrast to attached water or water of crystallization.

molecular sieve—adsorbent composition of aluminosilicates with pores of uniform molecular dimensions and used to selectively adsorb and gather sized molecules.

molecular weight—sum of the atomic weights of all the atoms in a molecule.

molecule—smallest particle of a chemical compound that can take part in a chemical reaction and is capable of independent existence with retention of all its chemical properties.

molten-metal corrosion—attack of a solid metal by contacting lower melting liquid metal; occasionally seen when lead is used as a heat-treating medium.

molten salt corrosion—corrosion of metals by molten or fused salts, e.g., fluorides, chlorides, nitrates, sulfates, hydroxides, and carbonates. Caused by metal dissolution or oxidation in the melt.

Monel—alloy of about 60 to 70% nickel and 25 to 35% copper with small quantities of iron, manganese, silicon, and carbon. Used to make acid-resisting equipment in the chemical industry.

monobasic acid—acid having only one acidic hydrogen atom in its molecule, e.g., HCl, HNO_3.

monofill—sludge-only landfill.

monolithic linings—linings of a single, thick (from $\frac{1}{2}$ inch to several inches) cementitious material applied by casting or troweling.

monolithic surfacing—chemically resistant resinous topping of a single composition and being greater than $\frac{1}{16}$ inch in thickness. Most commonly applied over concrete flooring or to line trenches, pits, and vessels.

monomer—single molecule or a substance consisting of single molecules. Term is mostly used in differentiation of dimer, trimer, etc., designating polymerized or associated molecules or substances composed of them. A relatively low-mass molecular structure that undergoes a polymerization reaction to form a polymer. Organic molecule or compound capable of polymerizing or linking together with itself or with other monomers to form a dimer, trimer, or polymer.

monomolecular layer—film on a solid or liquid surface that is only one molecule thick. Also called monolayer.

monovalent—having a valency of one.

montmorillonite—naturally occurring clay mineral (an aluminum magnesium silicate) forming the principal constituent of bentonite and fuller's earth.

mortar—mixture of gypsum plaster with aggregate or hydrate lime, or both, and water to produce a trowelable fluidity that becomes hard in place and used to bond units of masonry construction.

mortar test—procedure for determining the resistance of architectural metalwork and coated materials to damage from contact with wet mortar by applying a small amount of wet mortar to a test surface, allowing it to dry, and examining the surface after removal of the dried cement.

motion cell—corrosion cell developed by motion of the electrolyte past an electrode or motion of the electrode itself relative to another electrode.

mottled iron—cast iron containing a mixed structure of gray iron and white iron of variable proportions.

mpy—abbreviation for mils per year.

MSDS—abbreviation for Material Safety Data Sheet.

MTI—abbreviation for the Materials Technology Institute of the Chemical Process Industries, Inc.

mucilage—adhesive prepared from a gum and water.

mud—mixture of soil and water in a fluid or weakly solid state. Also a water- or oil-base drilling fluid whose properties have been altered by dissolved or suspended solids. Used for circulating out cuttings and for other functions while drilling a well.

mud cracking—cracking of an applied coating through to the substrate and occurring as soon as significant coating solvent or water evaporates.

multiphase coolant—engine coolant composed of immiscible liquids or undissolved solids, or both.

multiple bond—bond between two atoms that contains more than one pair of electrons.

municipal water—that water that has been found, treated, and distributed for industrial, commercial, and residential use.

Muntz metal—form of brass containing approximately 60% copper, 39% zinc, and small amounts of lead and iron. It is stronger than conventional brass and used primarily for condenser systems that employ fresh water as the coolant.

muriatic acid—commercial grade of hydrochloric acid (HCl).

MVT (MVTR)—*abbreviation for* **moisture vapor transmission rate**.

N

NACE International —abbreviation for the National Association of Corrosion Engineers International.

nailhead rusting—rust from iron and steel nails that penetrates or bleeds through a paint coating and stains the surrounding area.

naphtha—general term for flammable hydrocarbon solvents of both aliphatic and aromatic types derived by distillation from petroleum products or from coal tar. *See also* **high flash solvent naphtha** and **VM&P naphtha**.

naphthenic-based—petroleum products prepared from naphthenic-type crudes, i.e., crudes containing a high percentage of ring-type aliphatic hydrocarbon molecules.

NAS—abbreviation for the National Academy of Sciences.

natron—mineral form of hydrated sodium carbonate ($Na_2CO_3 \cdot H_2O$).

natural conversion coating—conversion coating, usually an oxide, formed on the surface of a metal on exposure in its natural environment in the presence of moisture. *See also* **conversion coating**.

natural gas—naturally occurring mixture of hydrocarbon and nonhydrocarbon gases found in porous geological formations beneath the surface of the earth, often in association with petroleum. The principal constituents of natural gas are methane (about 85%), ethane (up to about 10%), propane (about 3%), and butane. Carbon dioxide, nitrogen, oxygen, hydrogen sulfide and, sometimes, helium may also be present.

natural resin—*see* **resin, natural**.

natural rubber—product prepared by coagulating the latex of, primarily, the *Hevea brasiliensis* tree. Chemically, it is based on natural cis-polyisoprene, $[CH_2=C(CH_3)CH=CH_2]_n$. Depending on the degree of curing, natural rubber is classified as soft, medium, or hard.

natural weathering—exposure of materials to natural sunlight and climatic conditions.

naval brass—similar to Muntz metal but also containing 1% tin. *See also* **Muntz metal**.

naval stores—chemically reactive oils, resins, tars, and pitches derived from oleoresin from trees, principally the pine species.

NBS—abbreviation for the National Bureau of Standards.

NDT—*abbreviation for* **nondestructive testing**.

near-white blast—blast cleaning to a degree of cleanliness slightly less than white metal. This surface preparation is now termed very thorough blast cleaning by the ISO. *See also* **white blasting**.

neat cement—slurry composed of Portland cement and water.

neat petroleum—oil visibly free of contaminants.

neatsfoot oil—pale yellow animal oil made from the feet and shinbones of cattle.

neoprene—synthetic rubberlike film-former made by the polymerization of chloroprene (2-chloro-1, 3-butadiene). Used in place of natural rubber in applications requiring resistance to chemical attack.

Nernst equation—equation that expresses the exact electromotive force of a cell in terms of the activities of products and reactants of the cell. Makes it possible to determine the corrosion tendency of different metals and provides a means of computing the potential difference in bimetallic corrosion and also the maximum energy available from a galvanic cell. The equation is: $E_{eq} = E^0 + RT/nF \ln(ox/red)$ where E_{eq} is the equilibrium potential of the cell, E^0 is the standard potential of the cell, R is the perfect gas constant (1.987 cal/degree mole), T is the temperature in degrees kelvin, n is the number of electrons involved in the electrochemical reaction, F is the Faraday constant (23,060 calories/V equ), (ox) is the activity of the oxidized species, (red) is the activity of the reduced species, and E^0 is the standard potential of the electrode when the term (ox/red) is equal to 1.

Nernst layer (Nernst thickness)—diffusion layer or the thickness of this layer as given by the theory of Nernst. It is a hypothetical thickness that has been found to be 0.05 centimeter in many cases of unstirred aqueous electrolytes.

Nernst-Thompson Rule—solvents of high dielectric constant favor dissociation by reducing the electrostatic attraction between positive and negative ions, and conversely solvents of low dielectric constant have small dissociating influence on an electrolyte.

neutralization—chemical reaction that produces a resulting environment that is neither acidic nor alkaline. Also the reaction in which an acid and a base combine to yield a salt and, usually, water.

neutralizer—dilute alkaline solution used to treat sheet metal subsequent to the acid treatment of a pickling process.

neutralizing amines—chemicals added to the condensate of boiler systems to reduce metal loss from corrosion caused by carbon dioxide corrosion. They are volatile alkalizers that distill with the steam and neutralize acids generated by the breakdown of carbonates in the boiler water resulting in solution of carbon dioxide in the condensate. Commonly used neutralizing amines include ammonia, morpholine, diethylaminoethanol, cyclohexylamine, and dimethylisopropanolamine.

neutral oils—lubricating oils prepared by petroleum distillation and without chemical treatment.

neutral salt spray test—most common and widely used comparison accelerated corrosion test. See ASTM B 117 for details.

neutral solution—fluid environment containing an equal amount of hydrogen and hydroxyl ions, i.e., pH is 7.

neutron—elementary particle having no electric charge and a rest mass of 1.67×10^{-27} kilogram.

neutron embrittlement—metal or alloy embrittlement resulting from bombardment with neutrons, usually from exposure in the core of a reactor.

newton—symbol N. SI unit of force, being the force required to give a mass of one kilogram an acceleration of 1 ms^{-2}.

Newtonian fluid—true fluid that tends to exhibit constant viscosity at all rates of shear.

new work—applies to materials not already primed and painted and that require degreasing or removal of all oils, greases, waxes, and fatty compounds before coating. *See also* **old work**.

NF—abbreviation for the National Formulary.

Nichrome—trademark for a group of nickel-chromium alloys of nominal composition 80% nickel and 20% chromium used for wire in heating elements because of good resistance to oxidation and a high resistivity.

NIOSH—abbreviation for the National Institute for Occupational Safety and Health.

Ni-Resist—trademark for a class of austenitic, tough gray cast irons containing from 14 to 32% nickel and from 1.75 to 5.5% chromium, and with or without up to 7% copper.

nitriding—absorption of nitrogen atoms by a metal to remain either dissolved or to form metal nitrides. Also the introduction of nitrogen into the surface of a solid ferrous alloy by holding the alloy at a suitable temperature in contact with either ammonia or molten cyanide.

nitrile rubber—synthetic rubber which is a copolymer of 1-3 butadiene and propenonitrile. It has an outstanding resistance to oil and many solvents.

NLGI—abbreviation for the National Lubricating Grease Institute.

NLGI number—numerical scale for classifying the consistency of lubricating greases as developed by the National Lubricating Grease Institute (NLGI).

NMAB—abbreviation for the National Material Advisory Board.

NMR—*abbreviation for* **nuclear magnetic resonance**.

NO_x—term used to describe nitrogen oxides emitted to the atmosphere as pollutants which also contribute to atmospheric corrosivity. NO_x emissions originate mainly from different combustion processes, primarily road traffic and energy production and to a lesser degree from electric-discharge phenomena. In combustion processes, mostly NO is emitted, which is atmospherically oxidized to NO_2 or to a smaller extent, in the presence of water, to HNO_3.

noble—positive (increasingly oxidizing) direction of electrode potential. *Opposite of* **base metal**.

noble gases—group of monatomic gaseous elements forming group 0 (sometimes called group 8) of the periodic table and includes helium, neon, argon, krypton, xenon, and radon. Also called rare gases and inert gases.

noble metal—metal having a standard electrode potential that is more noble (positive) than that of hydrogen. The term is often used synonymously for precious metal when referring to such as platinum and gold. The noble metals include gold, silver, platinum, iridium, osmium, palladium, rhodium, and ruthenium.

noble potential—potential more cathodic (positive) than the standard hydrogen potential.

nodular graphite—graphite in the form of nodules or spheroids in iron castings.

nodule—rounded projection formed on a cathode during electrodeposition. Also little lumps.

noise—extraneous electronic signal that effects baseline stability.

nominal value—value assigned for the purpose of convenient designation and existing in name only.

nomograph—graph consisting of three coplanar curves, usually parallel straight lines, each graduated for a different variable so that a straight line cutting all three curves intersects the related values of each variable. Thus it is a graph representing numerical relationships. Also called a nomogram.

nonaqueous dispersion—nonaqueous analog of an aqueous dispersion where an organic solvent is the continuous phase.

nonaqueous phosphating—phosphate coating processes using organic solvents for phosphoric acid or phosphoric acid esters with or without organic film-forming resins or polymers present.

noncarbonate hardness—hardness in water caused by chlorides, sulfates, and nitrates of calcium and magnesium.

noncoating phosphate process—phosphate coatings that are much thinner than conventional iron phosphating coatings. Main solution constituents are alkali metal or ammonium dihydrogenphosphates. Also known as light weight iron phosphating or alkali metal phosphating.

noncombustible—that which does not burn. Preferred over the term incombustible.

noncondensibles—gases that are not condensed (liquefied) when associated water vapor is condensed in the same system.

nondestructive testing (NDT)—testing to detect internal and concealed defects in materials using techniques that do not damage or destroy the item being tested.

nondrying oil—oil that will not harden in air.

nonelectrolyte—substance that does not ionize or dissociate to produce ions when dissolved in solution.

nonferrous—containing no iron.

nonflammable—that which does not burn with a flame. Preferred over the term inflammable.

nonindustrial water—water used for nonindustrial facilities such as office buildings, department stores, hospitals, schools, apartments, residences, and various commercial enterprises.

nonionic—nonionized molecule in solution, or a colloidal particle without a surface charge.

nonionic surfactants—surfactants containing hydrophilic groups that do not ionize appreciably in aqueous solutions.

nonmetal—any of a class of chemical elements that are typically poor conductors of heat and electricity and that do not form positive ions. They are electronegative elements forming compounds that contain negative ions or covalent bonds, and their oxides are either neutral or acidic.

nonmetallic abrasives—naturally occurring, by-product, and manufactured abrasives that are not metallic.

nonmetallic inclusions—particles of impurities, e.g., sulfides, silicates, that are held mechanically or are formed during solidification or by subsequent reaction within solid metal.

non-Newtonian fluid—fluid that does not fall within Newton's mathematical definition of viscosity. *See also* **Newtonian fluid**.

nonoxidizing biocide—biocide whose effectiveness depends on some property other than the ability to oxidize organic matter, i.e., systemic poisons or surface active agents. *See also* **nonoxidizing toxicants**.

nonoxidizing toxicants—chemical compounds used in conjunction with oxidizing microbiocides for cooling water disinfection. Examples include chlorinated phenolics such as pentachlorophenate and various trichlorophenates, organo-tin compounds, quaternary ammonium salts, organo-sulfur compounds such as methylene bisthiocyanate, sulfones and thiones, and sodium dimethyldithiocarbamate.

nonpolar compound—compound that has covalent bonding only with no permanent dipole moment. *Opposite of* **polar compound**.

nonsoap grease—product similar to grease in appearance and consistency but containing only heavy residual oil stocks and mineral oil.

nonstoichiometric compound—chemical compound having a composition not in accord with the law of definite proportions. They occur among the binary compounds of Group 6b and especially among the intermetallic compounds. *Opposite of* **stoichiometric compound**.

nontoxic—not poisonous.

nonuniform corrosion—corrosion in small localized areas.

nonvolatile content—portion of a coating that does not evaporate during drying or curing and usually comprises the binder and pigment.

normalizing—heating a ferrous alloy to a temperature above the transformation range and then cooling in air to a temperature substantially below the transformation range.

normal salt—ionic compound containing neither replaceable hydrogen nor hydroxyl ions.

normal solution—solution containing one gram equivalent weight of the active reagent in one liter of the solution.

NPCA—abbreviation for the National Paint and Coatings Association.

NPDES—abbreviation for the U.S. National Pollutant Discharge Elimination System program.

NRCC—abbreviation for the National Research Council of Canada.

NTA—abbreviation for the sodium salt of nitrilotriacetic acid, a common chelant used in boiler water treatment to prevent the deposition of hardness and other heavy metals under boiler water conditions

NTP—abbreviation for the National Toxicology Program.

nuclear magnetic resonance (NMR)—a spectroscopic technique for chemical analysis and structure determination.

nuclear pressurized water reactor (PWR)—device utilizing the heat produced by nuclear reaction to generate electricity by raising steam to drive a steam turbine that turns a generator. The reactor has two major water systems: the primary loop to cool the reactor and to raise steam in the secondary loop, which drives the turbine.

nucleate boiling—formation of vapor bubbles (boiling) in a superheated liquid layer adjacent to a heat-transferring surface whose surface temperature is higher than the bulk liquid. The condition is encountered with engine coolants when an engine is being operated.

nucleate boiling cavitation corrosion—conjoint action of corrosion and impingement erosion resulting from nucleate boiling.

nucleon—a proton or neutron.

nucleus—small, dense, positively charged region in the center of an atom containing the protons and neutrons.

nuclide—species of atom distinguished by the constitution of its nucleus.

NVM—abbreviation for nonvolatile matter.

nylon resins—generic name for all synthetic polyamides. Synthetic resins composed principally of long-chain polymeric amide that has recurring amide groups ($-CONH_2$) as an integral part of the main polymer chain.

OCCA—abbreviation for the Oil and Colour Chemists' Association (British).

occlusion—absorption process by which one solid material adheres strongly to another, sometimes occurring by coprecipitation. Also the trapping of undissolved gas in a solid during solidification.

ocean water—entire body of seawater that covers about 72% of the earth's surface.

ochre—yellow or red mineral form of iron(III) oxide (Fe_2O_3) and used as a pigment.

ODS—abbreviation for ozone-depleting substances, e.g., chlorofluorocarbons (CFC-113) and methyl chloroform (MFC) or (1,1,1-trichloroethane).

OE or OEM—abbreviation for original equipment or original equipment manufacturer.

ohm—derived SI unit of electrical resistance; being the resistance between two points on a conductor when a constant potential difference of one volt applied between these points produces a current of one ampere in the conductor.

ohmic corrosion inhibitors—corrosion inhibitors that increase the ohmic resistance of the electrolyte circuit by selectively forming a film, usually of a microinch or more, on the metal surface either on its anodic or cathodic areas or on both areas.

ohmic drop—type of overpotential resulting from the resistance of the electrolyte to the flow of electric current. This flow depends on the number of ions available for charge transport and their ability to move as influenced by interactions with other solution particles.

ohmmeter—direct-reading instrument for measuring the value of a resistance in ohms.

Ohm's law—$I = E/R$ where I is the current in amperes, E is the potential in volts, and R is the resistance in ohms.

oil—nonspecific term for any of numerous viscous liquids that are generally unctuous, combustible, immiscible with water and soluble in various organic solvents. Natural plant and animal oils are either volatile mixtures of terpenes and simple esters, e.g., essential oils, or are glycerides of fatty acids. Mineral oils are mixtures of hydrocarbons, e.g., petroleum. Oils may also be of synthetic origin.

oil ash corrosion—corrosion wastage of secondary superheater and reheater tube sections of boilers using residual oil or bunker "C" oil high in vanadium content as the fuel. Molten ash deposits from this fuel dissolve protective iron oxide layers on the tube surfaces, act as an oxidation catalyst, and allow oxygen and other combustion gases to diffuse rapidly to the metal surface. The corrosion is accelerated by the added presence of sodium, chlorine, and sulfur in the ash.

oil-base mud—drilling fluid where oil is the continuous phase and water the dispersed phase.

oil dehydration—separation of produced water from crude oil. Steps include free-water separation on standing followed by breaking remaining emulsified water usually assisted by heat, electric energy, or both.

oil fouling—presence of lubricating oil in the cooling system of an engine.

oil of vitriol—sulfuric acid (H_2SO_4).

oil/gas well corrosion inhibitors—corrosion inhibitors applied to natural oil/gas wells to counteract their natural corrosivity resulting from the presence of water, salt, CO_2, and H_2S in the hydrocarbon mix. The inhibitors are most commonly classified as to their solubility and dispersibility in the two phases (water/oil-gas) of the well. This classification describes them as being either (1) oil and water soluble, (2) oil soluble and water dispersible, (3) oil insoluble and water soluble, (4) oil and water insoluble, and (5) volatile inhibitors. The most commonly employed inhibitors are organic nitrogen molecules whose molecular weights exceed 200.

oil length—ratio of oil to resin in a medium. May be expressed in terms of parts by weight of oil to one part by weight of resin or in gallons of oil per 100 pounds of resin or as the percentage of oil by weight in the resin.

oil paint—paint composition containing a drying oil, oil varnish, or oil-modified resin as the vehicle. A common, but technically incorrect, definition is that it is any paint soluble in organic solvents.

oil sand—*see* **tar sands**.

oil shale—carbonaceous sedimentary rock containing organic matter called kerogen having an approximate composition of $(C_6H_8O)_n$ where *n* is a large number. When heated in the absence of air, the kerogen breaks down, forming hydrocarbon oils similar to petroleum.

oil spot test—test for metal surface cleanliness. A droplet of degreasing solvent is applied to the metal surface and then evaporated. Formation of a ring suggests the presence of surface oil on the metal surface.

oil well—hole dug or drilled in the earth from which petroleum flows or is pumped. Categorized as either a flowing well, which has a natural flow of hydrocarbons and water to the surface, or an artificially lifted well, which requires some form of pump to recover the petroleum.

oil-well pumps—lifting pumps that are used for oil recovery from a reservoir when pressure in the well has fallen to the point where a well will not produce by natural energy. The pumps are of three general types: (1) pumps located at the bottom of the hole and run by a string of rods, (2) pumps at the bottom of the hole run by high-pressure liquids, and (3) bottom-hole centrifugal pumps.

old work—applies to materials that have been painted or primed before and that require stripping and rust removal before being treated as new work ready for coating. *See also* **new work**.

olefin—family of unsaturated hydrocarbons with the formula C_nH_n, e.g., ethylene, propylene. *See also* **alkenes**.

olefin plastic—plastic based on polymers produced by the polymerization of olefins or co-polymerization of olefins with other monomers with the olefin being at least 50 mass%.

oleophilic—liking oil; wettable with oils.

oleoresinous—coating basis made combining an oil and resin, such as in a varnish.

oleum—disulfuric(VI) acid (pyrosulfuric acid), $H_2S_2O_7$, formed by dissolving sulfur trioxide in concentrated sulfuric acid. Widely used in the sulfonation of organic compounds. Also called fuming sulfuric acid.

oligomer—polymer of only a few monomer units, such as a dimer, trimer, tetramer, or their mixtures.

oligotrophic—natural body of water with a poor supply of nutrients and low rate of formation of organic matter by photosynthesis. *Contrast with* **eutrophic**.

once-through cooling water system—system whose cooling water passes through a heat exchanger only once before its discharge.

Oncor M-50—trademark for an anticorrosive pigment chemically described as basic lead silicochromate.

onium ion—ion formed by adding a proton to a neutral molecule, e.g., hydronium ion (H_3O^+), ammonium ion (NH_4^+).

open-circuit potential—measured potential of a cell in which no current flows. *Also called* **rest electrode potential**. *See also* **corrosion potential** and **electrode potential**.

open recirculating cooling water system—characteristically removes its heat by evaporative cooling, e.g., spray ponds and cooling towers.

operating cycle—ion-exchange process term consisting of the combination of backwash, regeneration, rinse, and service run.

optical microscopy—use of magnifications of about 10 to 1000× to examine a surface illuminated with visible light.

ore—mineral containing useful metallic substances that can be extracted.

organic—being composed of hydrocarbons or their derivatives, or matter of plant or animal origin. *Contrast with* **inorganic**.

organic acid—chemical compound with one or more carboxyl radicals (–COOH) in its structure.

organic acid cleaning—use of organic acids for metal cleaning applications. *See also* **acid cleaning**. Primary organic acids used include acetic acid, citric acid, EDTA, formic acid, gluconic acid, and hydroxyacetic acid.

organic chemistry—branch of chemistry concerned with compounds of carbon.

organic compound—any carbon compound except for those that exhibit properties of inorganic compounds.

organic solvent—organic material that is liquid at standard conditions and that is used as a dissolver, diluent, viscosity reducer, or cleaning agent.

organic zinc-rich paint—paint or coating containing zinc powder pigment and an organic resin binder.

organo-—used as a prefix before the name of an element indicating compounds of that element containing an organic group or groups where the element is bound to a carbon atom.

organometallic compound (metal-organic compound)—compound in which one or more hydrogen atoms of an organic compound have been replaced by a metallic atom or atoms, usually with the establishment of a valence bond between the metal atom and a carbon atom. A metallic salt of an organic acid is not classified as an organometallic compound.

organosol—suspension of a finely divided high-molecular-weight resin, with little or no plasticizer present, in a volatile organic liquid. It is applied as a coating mixture where the resin is solvated and then fuses to itself and the coated surface. *See also* **plastisol**.

ORP—*abbreviation for* **oxidation-reduction potential** or **redox potential**.

orthophosphoric acid—compound of the composition H_3PO_4.

OSHA—abbreviation for the U.S. Occupational Safety and Health Administration.

osmosis—passage of water through a semipermeable membrane separating two solutions of different concentrations. The water passes into the more concentrated solution.

osmotic pressure—pressure that is required to just stop the movement of solvent across a semipermeable membrane (osmosis) from pure water to a solution.

outgassing—evolution of gas from a material in a vacuum.

overpickling—porosity of the transverse surfaces and a roughening of the whole surface of pickled metal, accompanied by its discoloration and decrease in size and weight, resulting from overly long immersion of the metal in the pickling bath.

overpotential—potential change of a chemically reversible electrode when its double layer is crossed by an electric current. Also the potential difference between a working electrode and its equilibrium potential. Also a potential that must be applied in an electrolytic cell in addition to the theoretical potential required to liberate a given substance at an electrode. The magnitude of overpotential is proportional to current density and to the resistance offered by the system to the flow of electric current. The total overpotential is the sum of all the types of overpotentials present. These types include: (a) ohmic drop overpotential resulting from the resistance of the electrolyte to the flow of electric current; (b) charge-transfer overpotential resulting from the resistance offered to the transfer of electric charges through the double layer; (c) concentration or diffusion overpotential producing a shift in the electrode potential resulting from a charge being generated in the reactant or product concentration by the electric current at the metal-electrolyte interface; (d) crystallization overpotential that is linked to the energy required by a just-reduced atom adsorbed at the metallic surface to occupy a site on the metallic lattice; (e) reaction overpotential resulting from chemical reactions occurring at the interface or in the electrolyte and affecting the concentration of the reacting species at the metal-solution interface; and (f) adsorption overpotential resulting from the energy required for adsorption of reacting species.

overvoltage—difference in electrode potential when a current is flowing versus when there is no current flow. *Also known as* **polarization**. Also is an example of overpotential.

oxidant—*see* **oxidizing agent**.

oxidation—reaction in which electrons are removed from a reactant. A reaction in which there is an increase in valence resulting from a loss of electrons. *Contrast with* **reduction**. Also refers to the corrosion of a metal that is exposed to an oxidizing gas at elevated temperature.

oxidation inhibitor—substance added to materials to improve their resistance to deleterious attack in an oxidizing environment.

oxidation number (oxidation state)—measure of the electron control that an atom has in a compound compared to the atom in the pure element. An oxidation number consists of two parts: (1) its sign, which indicates whether this control has increased (negative) or decreased (positive); (2) its value, which gives the number of electrons over which control has changed. This change in electron control may be complete (in ionic compounds) or partial (in covalent compounds). Oxidation is a reaction involving an increase in oxidation number, and reduction involves a decrease. *See also* **oxidation-reduction**.

oxidation potential—potential drop involved in the oxidation, i.e., ionization, of a neutral atom to a cation, of an anion to a neutral atom, or of an ion to a more highly charged state.

oxidation-reduction—oxidation is the loss of electrons by a chemical species and reduction is the gain of electrons. The combined term oxidation-reduction is used to describe the simultaneous occurrence of oxidation and reduction because whenever one occurs the other must also in order to maintain electrical neutrality of the system.

oxidation-reduction potential—relative potential of an electrochemical reaction under equilibrium or nonreacting (zero current flow) conditions. *Also known as* **redox potential**, half-cell potential, and **electromotive force** or emf (EMF).

oxidation state—*see* **oxidation number**.

oxide conversion coating—oxide-type conversion coating chemically produced on iron, steel, stainless steel, aluminum, copper and copper alloys, zinc, and cadmium surfaces to provide corrosion resistance, abrasion resistance, or color to the base metal. *See also* **conversion coating**.

oxides—binary compounds formed between elements and oxygen.

oxidized—undergoing a loss of one or more electrons. Contrast with reduced. Also used to describe a metal surface that has been dulled or darkened by surface treatment or natural oxidation.

oxidized surface—refers to a steel surface having a thin, tightly adhering, oxidized skin from straw to blue in color.

oxidizing acid—acid that can act as a strong oxidizing agent as well as an acid, e.g., nitric acid (HNO_3).

oxidizing agent—compound that causes oxidation and is itself reduced. Also known as an oxidant.

oxidizing biocide—biocide whose effectiveness depends on its ability to oxidize and thus destroy organic matter, e.g., chlorine, bromine, and ozone.

oxidizing power (potential)—term relating to the ability to remove or add electrons from a metal so as to oxidize or reduce the surface. This potential can be applied by an external voltage source, by galvanic coupling of different metals, or by solution constituents.

oximes—organic compounds containing the group >C=NOH and formed by reaction of an aldehyde or ketone with hydroxylamine (H_2NOH). Used to transform rusty surfaces to a metal organic complex by chemical reaction. The converted surface is dark colored, stable, insoluble, and water repellent.

oxonium ion—ion of the type R_3O^+ where R indicates hydrogen or an organic group. The specialized term hydroxonium ion (H_3O^+), is used in place of oxonium ion when acids dissociate in water.

oxyacid—inorganic acid that contains oxygen, e.g., HNO_3, H_2SO_4.

oxygen concentration cell—galvanic cell caused by a difference in oxygen concentration at two points on a metal surface. *Also known as* **differential aeration cell**.

oxygen demand—amount of oxygen required for the oxidation of waterborne organic and inorganic matter.

oxygen scavengers—chemical additives to aqueous systems that inhibit corrosion by preventing the cathodic depolarization caused by oxygen. They do this by reacting with dissolved oxygen to remove it as a possible corrodent. Examples include sodium sulfite, sulfur dioxide, and hydrazine.

ozone—powerful oxidizing allotropic form of the element oxygen. The ozone molecule consists of three atoms of oxygen (O_3).

ozone cracking—fissures developed in the surface of rubber under a tensile strain upon exposure to ozone.

P Q

pack cementation—high-temperature diffusion of certain materials, e.g., carbon, aluminum, chromium, boron, zinc, etc., into the surface of a metal to produce a surface product entirely different from the metal.

packer fluid—any fluid placed in the annulus between the tubing and casing above a packer of an oil and gas well. Its primary purpose is to maintain hydrostatic pressure on the casing interior, tubing exterior, and the packer to prevent development of potentially dangerous pressure on the casing. Packer fluids must also protect the outside of the tubing and inside of the casing against corrosion during the productive life of the well. Fluids utilized for packer fluids include water-base drilling muds, clear solutions (fresh water and sodium chloride and calcium chloride solutions), and oil muds.

packing—inert fill material in a vessel or chamber used to provide a large surface area per unit volume and used in scrubbers. *See also* **scrubber**.

packout rust formation—water draining condition that can occur in steel bolted structures having excessive spacing due to loose bolting and where retained moisture can initiate corrosion with the expanded rust product causing distortion of the structure.

paint—pigmented coating. Also a mixture composed of solid coloring matter suspended in a liquid medium and applied as an adhering coating to various types of surfaces. Also material that, when applied as a liquid to a surface, forms a solid film for the purpose of protection or decoration. Also a classification employed to distinguish pigmented drying oil coatings (paints) from synthetic enamels and lacquers. The most commonly utilized paints and their vehicles are: oil paints (boiled linseed oil, soybean oil, or similar materials); phenolic resin paints (phenolic resins with vegetable oils), alkyd resin paints (phthalic acid with polyhydric alcohols, e.g., glycerol and vegetable oil or fatty acid), melamine resin paints (melamine and alkyd resin combinations), vinyl resin paints (copolymers of chlorinated vinyl and other resins), epoxy resin paints (epoxy resins), polyurethane resin paints (polyol resins cured with isocyanate groups), acrylic resin paints (acrylic resins), silicone alkyd modified resin paints (silicone with alkyd resins), tar-epoxy resin paints (coal tar with epoxy resins), and polyester resin paints (polyhydric alcohol with polybasic acid). The protective properties of paints include defense against air, water, sunlight, and chemicals, as well as imparting a coating of some hardness and abrasion resistance.

painting system—term that includes the surface preparation, paint materials, their application, inspection, and the safety aspects of painting.

palming—mitten application of a paint coating.

paper chromatography—technique for analyzing mixtures by chromatography in which the stationary phase is absorbent paper.

paraffin—any solid hydrocarbon of the methane series boiling above 300 °C (572 °F). *See also* **alkanes**.

paraffin-based—petroleum products prepared from paraffin-type crudes, which are crudes containing a high percentage of straight chain aliphatic or paraffinic hydrocarbon molecules.

paraffinic oil—mineral oil derived from crudes having substantial contents of naturally occurring wax.

paraffin oil—oil, either pressed or dry-distilled, from paraffin distillate.

paraffin wax—inert hydrocarbon wax derivative of crude petroleum. Also defined as a distillate, of mainly *n*-paraffins, with a molecular weight of 225 to 450 and a setting point of 32 to 71 °C (90 to 160 °F).

parasitic corrosion—corrosion of the aluminum anode in aluminum/air battery systems employing an alkaline electrolyte.

Parkerized—trademark for the product of a process for the conversion coating of iron and steel with a dark, corrosion-resistant and generally protective coating that also improves the bonding of subsequent paint and lacquer films. The ferrous body is immersed in a boiling aqueous solution of manganese dihydrogen phosphate and may optionally also have applied a coating of paraffin oil if it is not to be painted. *See also* **Bonderize**.

partial annealing—imprecise term denoting a treatment given cold-worked metal to reduce its strength to a controlled level or to effect stress relief.

partial pressure—pressure exerted by an individual gas in a gaseous mixture.

particulate—any suspended or undissolved particle in a liquid or gas.

particulate matter—nonliquid matter, exclusive of gases, that is heterogeneously dispersed in water or some other liquid.

parting corrosion—selective corrosion or leaching of a component from an alloy, such as the parting of zinc from brass to leave a copper residue (dezincification). *Also known as* **dealloying corrosion**.

parting limit—minimum concentration of a more noble component in an alloy, above which concentration parting does not occur in a specific environment.

parts per million (ppm)—measure of proportion by weight; being the number of weight parts of a substance per million weight parts of the total composition.

pascal—Symbol Pa. SI unit of pressure equal to one newton per square meter.

passivating—process for creating passivity.

passivation—process to render the surface condition of a metal to a much less reactive state. *Contrast with* **activation**.

passivation inhibitor—corrosion inhibitor that forms a protective oxide film on the metal surface. Examples include chromate, nitrite, molybdate, and tungstate, which promote passivation by increasing the potential of iron or steel surfaces. *See also* **adsorption inhibitor**, **corrosion inhibitor**, and **precipitation inhibitor**.

passivation potential—*see* **primary passive potential**.

passivator—chemical corrosion inhibitor that appreciably changes the potential of a metal to a more noble (positive) value.

passive—state of a metal surface characterized by low corrosion rates in a potential region that is strongly oxidizing to the metal. *Contrast with* **active**.

passive-active cell—corrosion cell in which the anode is a metal in the active state and the cathode is the same metal in the passive state.

passive metal—metal that substantially resists corrosion in a given environment resulting from marked anodic polarization. Also a metal that substantially resists corrosion in a given environment despite a marked thermodynamic tendency to react.

passivity—loss of chemical reactivity by a metal or alloy under specific environmental conditions. Also the condition in which a metal, because of an impervious covering of an oxide or other compound, has a potential much more noble than the bare metal.

pasteurization—process of destroying most disease-producing microorganisms and for limiting fermentation in milk, beer, and other liquids by their partial or complete sterilization.

patenting—heating a ferrous alloy above the critical temperature range then cooling below that range in air or some molten medium at a temperature usually between 425 and 555 °C (800 and 1050 °F) as a treatment to precede wire drawing.

pathogens—disease-producing microbes.

patina—greenish corrosion coating that forms on copper and some copper alloys, consisting mainly of copper sulfates, carbonates, and chlorides, after exposure to the atmosphere for some time.

PCB—abbreviation for polychlorinated biphenyls.

PE—*abbreviation for* **polyethylene**.

pearl ash—commercial potassium carbonate.

pearlite—metastable lamellar aggregate of ferrite and cementite resulting from the transformation of austenite at temperatures above the bainite range.

pebbly—having a rough and irregularly indented surface.

peeling—spontaneous detachment in ribbons or sheets of a paint or an electrodeposited metallic coating from a basis metal or undercoat. Also the separation of any coating from the substrate initiating at an edge of or cut in the coating and causing an exposure of the substrate.

peening—use of metallic shot to impart residual compressive stresses to improve the fatigue properties of metal products, and to minimize intergranular and stress-corrosion cracking of alloyed metal products.

peen plating—*see* **mechanical plating**.

PEI—abbreviation for the Porcelain Enamel Institute, Inc.

PEL—*abbreviation for* **permissible exposure limits**.

penetrant—solution of dye, either visible or fluorescent, capable of entering discontinuities open to the surface. *See also* **black light**.

penetration number—depth, in tenths of a millimeter, that a standard cone penetrates a semisolid sample under specified conditions. ASTM Designation D 217.

per-—prefix indicating a chemical compound containing an excess of an element, e.g., peroxide.

perched aquifer—water-bearing geologic formation in which a limited layer of impermeable material occurs above the main water table allowing a thin zone of saturation to form above it.

perfect gas—*see* **ideal gas**.

perfluoroalkoxytetrafluoroethylene (PFA)—copolymer of tetrafluoroethylene and a perfluorovinyl ether having a service temperature to 195 to 230 °C (425 to 450 °F) and is similar in corrosion resistance to PTFE.

period—horizontal row of elements on the periodic table.

periodic chart—arrangement of the elements in order of increasing proton (atomic) number that illustrates the repetition or periodicity of key characteristics.

periodic law—principle that the physical and chemical properties of elements are a periodic function of their proton (atomic) number.

periodic table—table of elements arranged in order of increasing proton number to show the similarities of chemical elements with related electronic configurations.

perlite—small, grayish, natural volcanic siliceous glass spherules similar to obsidian. In a fluffy heat-expanded form, it is used as a lightweight aggregate in plaster and concrete and in thermal and acoustic insulation.

perm—unit of measurement of water vapor transmission used to express the resistance of a material to the penetration of moisture. The U.S. perm unit is equal to 1 grain of water/hour/ft^2/in. Hg pressure.

permanence—organic coating term defined as the property of being permanent, i.e., lasting indefinitely without change.

permanent hardness—water containing dissolved calcium sulfate ($CaSO_4$) or calcium fluoride (CaF_2). These compounds cannot be removed from water by boiling, hence the name permanent hardness.

permeability—unit of permeance, i.e., water vapor passage through a film.

permeation—process whereby permeant (gas, liquid, vapor, or ions) in contact with a coating is absorbed, moves through the coating, and attains a state of dynamic equilibrium.

permissible exposure limits (PEL)—as defined by OSHA. It is the concentration of a contaminant in air to which workers can be exposed for a normal 8-hour day, 40-hour work week without ill effects as defined by OSHA.

permittivity—*see* **dielectric constant**.

Permutit—trademark for certain zeolites used for water softening.

peroxide—compound that contains an oxygen-oxygen single bond, e.g., H_2O_2.

Perspex—trademark for a form of polymethylmethacrylate.

perspiration—saline moisture excreted through the pores of the skin by the sweat glands. *Also known as* **sweat**. Perspiration may range from an acidic pH about 4.5 to an alkaline pH about 8.7.

pesticide—substance used to kill pests, especially insects and rodents.

PET—*abbreviation for* **polyterephthalate**.

petri dish—shallow, circular flat-bottomed dish of plastic or glass and having a fitting lid.

petrochemicals—organic chemicals obtained or derived from petroleum or natural gas.

petrolatum—residual mixture of semisolid hydrocarbons composed of microcrystalline and amorphous waxes and having an oily or greaselike characteristic.

petrolatum tape coating—coating tape consisting of a synthetic fiber fabric impregnated with a uniform thickness of petrolatum.

petrolatum wax (mineral jelly)—mixture of microcrystalline wax and about 10% mineral oil giving a white amorphous translucent wax finding use as a temporary protective.

petroleum—naturally occurring crude oil that consists chiefly of hydrocarbons with some other elements, such as sulfur, oxygen, and nitrogen. Crude oil is classified according to its base: paraffin-base, which is high in wax and lubricating oil fractions; asphalt-base, which is high in pitch, asphalt, and heavy fuel oil; mixed-base, having characteristics between paraffin- and asphalt-bases; and aromatic- or naphthenic-base, which is high in low molecular weight aromatic compounds and naphthene.

petroleum ether—colorless, volatile flammable mixture of hydrocarbons, mainly pentane and hexane, boiling in the range 30 to 70 °C (86 to 158 °F) and used as a solvent. Not an ether.

petroleum jelly—semisolid mixture of hydrocarbons having utility as a temporary corrosion protective film.

petroleum pitch—dark brown to black, predominantly aromatic, solid cementatious material obtained from petroleum, its fractions or residuals.

petroleum spirits (mineral spirits)—straight-run or blended petroleum naphthas with a boiling range between 150 to 205 °C (300 to 400 °F) and used as thinners for paints, enamels, and varnishes.

pewter—alloy of about 63% tin and lead with small amounts of copper or antimony.

PFA—*abbreviation for* **perfluoroalkoxytetrafluoroethylene**.

pH—measure of the acidity or alkalinity of a solution. A means of expressing hydrogen ion concentration in terms of the powers of 10. A value of 7 is neutral; lower numbers are acid; and larger numbers are alkaline. pH is the negative logarithm of the hydrogen ion concentration of a solution. The term pH scale derives from the French "pouvoir hydrogene" (hydrogen power).

pH_s—pH of saturation, i.e., pH at which water is in equilibrium with solid $CaCO_3$. The $CaCO_3$ has neither a tendency to dissolve nor precipitate. Term is used in the calculation of the Langelier saturation index and Ryzner stability index. *See also* **saturation index, Langelier saturation index,** and **Ryzner stability index**.

pH scale—logarithmic scale for expressing the acidity or alkalinity of a solution. *See also* **pH**.

phase—chemically homogeneous, distinguishable portion of a material system existing in the physical state. A phase can also be a homogeneous part of a heterogeneous mixture.

phase diagram—*see* **constitutional diagram**.

phenol—white, crystalline organic solid (C_6H_5OH). Also known as carbolic acid. Also any organic compound that contains an –OH group bonded to a benzene ring.

phenolic compounds—hydroxy derivatives of benzene and its condensed nuclei. Synthetic resin produced by the condensation of an aromatic alcohol with an aldehyde.

phenolic resins—resins made from the reaction of phenols and aldehydes.

phenolphthalein—water-soluble color indicator of pH. In water, its color is red at a pH value of 8.3 or higher and colorless at lower pH values.

pholads—small rock- and wood-boring clams.

phosphate-coated (phosphatized)—iron or steel chemically treated to provide a gray, somewhat protective ferric phosphate coating which also serves as a paint base. *See also* **phosphating**.

phosphate esters—principally, amine phosphates characterized by carbon to oxygen to phosphorus bonding and used for both corrosion and scale control and for metallic oxide sequestration in cooling water systems.

phosphate rustproofing—use of heavy phosphate coating in conjunction with an oil or wax to protect steel from rusting. *See also* **phosphating**.

phosphates—chemical compounds that are salts of phosphoric acid. Also salts based on phosphorus (V) oxo acids. There are three general forms of phosphates: orthophosphates (PO_4^{3-}), condensed phosphates, sometimes called acid hydrolyzable phosphates,

phosphating

which occur in such compounds as metaphosphates, pyrophosphates, and polyphosphates; and organic phosphates, sometimes referred to as organically bound phosphates.

phosphating—process for forming an adherent phosphate coating on a metal by immersion in a suitable aqueous phosphate solution and phosphoric acid. Also called phosphatizing. *See also* **conversion coating**.

phosphating lacquer—temporary protection coating formed from composition consisting of phosphoric acid in a polar solvent also containing an organic resin such as polyvinylbutyral or a phenolic. Not a true wash primer.

phosphonates—organic polymers whose carbon to phosphorus to oxygen bonding is more resistant to hydrolysis or cleavage than the COP bond in phosphate esters. The applications for the two, however, are similar. *See also* **phosphate esters**.

phosphonic acid—dibasic acid (H_3PO_3), more correctly written as $(HO)_2HPO$, that gives rise to the ions $H_2PO_3^-$ and HPO_3^{2-} and is moderately reducing. Also known as phosphorus acid.

phosphor bronze—copper alloy nominally consisting of 90 to 95% copper, 5 to 10% tin, and 0.25% phosphorus.

phosphorus acid—*see* **phosphonic acid**.

photochemically reactive organic material—any organic material that will react with oxygen, ozone, or free radicals generated by the action of sunlight on components in the atmosphere to result in a deterioration of its properties.

photochemical oxidants—air pollutants formed by the action of sunlight on oxides of nitrogen and on hydrocarbons.

photochemical reaction—chemical reaction caused by light or ultraviolet radiation.

photochemical smog—air pollution caused by not one pollutant but by chemical reactions of various pollutants emitted from different sources.

photoelectric cell (photovoltaic cell)—corrosion cell developed when one electrode is in the light and the other in the dark.

photoelectric effect—liberation of electrons from a substance exposed to electromagnetic radiation.

photolysis—chemical reaction produced by exposure to light or ultraviolet radiation.

phthalate esters—group of widely used plasticizers for paints and plastics made by the reaction of an alcohol with phthalic anhydride ($C_6H_4(CO)_2O$).

physical adsorption—*see* **physisorption** and **van der Waals force**.

physical change—change in the physical properties of a substance with no change in composition.

physical property—property associated with an individual substance that can be described without referring to any other substance, e.g., color, size, mass, density, strength, etc.

physical vapor deposition (PVD)—coating process whereby the work is heated and rotated on a spindle above the streaming vapor that is generated by melting and evaporating a coating material source bar with a focused electron beam in an evacuated chamber.

physisorption—binding of an adsorbate to the surface of a solid by forces whose energy levels approximate those of condensation. *Contrast with* **chemisorption**.

pickle—solution, usually an acid, used to remove mill scale or other corrosion products from a metal. Some base metal may also be removed in the pickling process.

pickle liquor—acid used in treating steel for removal of oxide scale.

pickling smut—residue remaining after and as the result of pickling. In the case of pickled steel or cast iron, smut is mostly ferric carbide with some iron oxide and fine carbon. Usually removed from the surface by ultrasonic cleaning or alkaline electrocleaning.

pig—generally geometrically conforming device that is forced through a pipeline with a liquid or gas to clean debris, etc., from the inside of the pipe. It can also be used to separate different types of liquids in a pipeline. The pig itself is often a flexible bullet-shaped foam cylinder that is propelled through the pipeline with water.

pig iron—impure form of iron (92 to 95% with about 4% carbon) produced by a blast furnace and cast into pigs (blocks) for converting at a later time into cast iron, steel, etc.

pigment—dry, insoluble matter used to impart color, opacity, bulking, or protective functions in coatings or plastics.

pigment volume concentration (PVC)—percent, by volume, of pigment in the nonvolatile portion of a paint.

pine oil—colorless to amber-colored volatile oil with characteristic odor derived from pine woods. It consists mainly of isomeric tertiary and secondary, cyclic terpene alcohols.

pine tar oil—blackish-brown oil obtained by condensing vapors from resinous pine wood undergoing destructive distillation. Its principal constituents are: turpentine, resin, guaiacol, creosol, methylcreosol, phenol, phlorol, toluene, xylene, and other hydrocarbons.

pinhole—tiny hole in a film, foil, or laminate, comparable in size to one made by a pin.

pinpoint rusting—coating term usually applied to the onset of spot rusting where a coating is thin and not fully protective or has been in service for a long time and is nearing the end of its useful life.

Pirani gage—instrument used to measure low pressures (100 to 0.01 Pa).

pit—small depression or cavity produced in a metal surface by corrosion.

pitch—black or dark brown residue resulting from the distillation of coal tar, wood tar, or petroleum (bitumen). Also sometimes used to describe naturally occurring asphalt. Used as a binding agent, fuel, and for waterproofing.

pith—soft, spongelike wood in the center of trees.

pitting—corrosion of a metal surface confined to a point or small area that takes the form of a pit or cavity. A form of highly localized corrosion.

pitting factor—depth of the deepest pit divided by the average penetration as calculated from weight loss.

PIXE—abbreviation for proton-induced X-ray emission.

plankton—minute organisms that drift or float passively with the current in sea or lake.

plasma—partially or totally ionized gas or vapor. Also the fluid portion of blood.

plasma deposition—method of applying powder to a surface using compressed gas to melt powder in a plasma heat source before impinging it on the surface.

plasma enhancement—used to describe the various methods of increasing the number of ions, electrons, fragmented molecules, and excited neutrons in a gas or vapor.

plasma spraying—thermal spraying process in which the coating material is melted with heat from a plasma torch that generates a nontransferred arc and the molten coating material is propelled against the base metal by the hot ionized gas issuing from the torch.

plaster—plastic system of solid particles that are free to move in a liquid. Usually refers to a mixture of Portland cement, lime, or gypsum, with water and sand that sets to form a hard surface.

plaster of Paris—plaster consisting of calcium sulfate hemihydrate and water, which sets to gypsum (calcium sulfate dihydrate). *See also* **calcined gypsum** and **gypsum cement**.

plastic—any polymeric material, either thermoplastic or thermosetting, of high molecular weight that can be molded, cast, extruded, drawn, laminated, or otherwise fabricated into objects, powders, beads, films, filaments, fibers, or other shapes. Rubber, textiles, adhesives, and paint, which may in some cases meet this definition, are not considered plastics. The term plastic is also used to describe material capable of being deformed continuously and permanently in any direction without rupture under a stress exceeding the yield value.

plastic deformation—permanent (inelastic) distortion of metals under applied stresses that strain the material beyond its elastic limit. *Opposite of* **elastic deformation**.

plastic flow—deformation of a plastic material beyond the point of recovery, accompanied by continuing deformation with no further increase in stress.

plastic fluid—non-Newtonian fluid in which the shear force is not proportional to the shear rate and a definite pressure is required to start and maintain movement of the fluid.

plasticity—property of solids that causes them to change permanently in size or shape as a result of the application of a stress in excess of a certain value, called the yield point.

plasticize—to soften a material making it plastic or moldable by means of either incorporating a plasticizer or the application of heat.

plasticizer—substance incorporated into a material to increase its workability, flexibility, or distensibility.

plastic-lined steel pipe—steel pipe that is plastic lined and used for a range of corrosive flow media. The steel shell provides mechanical strength, while the plastic liner provides corrosion resistance. Most commonly used liners are polypropylene, polyvinylidene chloride, polyvinylidene fluoride, fluorinated ethylene propylene copolymer, and polytetrafluoroethylene.

plastic media blasting (PMB)—abrasive blasting using recyclable plastic particles pneumatically applied at relatively low pressures. The plastic particles vary in hardness from 3 to 4 Mohs compared with most abrasives at about 7 Mohs. Used to remove paint coatings without harming sensitive substrates such as copper, brass, aluminum, magnesium, fiberglass, advanced composites, engineered plastics, etc.

plastisol—suspension of finely divided polymer in a plasticizer with no solvents. The polymer becomes dissolved in the plasticizer to form a homogeneous plasticized coating when applied as a coating and heated. Organosols differ in that they contain volatile thinners and little or no plasticizer.

plating—deposition of a thin layer of coating of metal. *Also called* **electrodeposition**.

plating range—current density range over which a satisfactory electroplate can be deposited.

platinum black—vacuum evaporated finely divided platinum metal that is black in appearance.

platinum metals—group VIII metal series consisting of platinum, ruthenium, rhodium, palladium, osmium, and indium, all of which are quite resistant to corrosion.

plug-type dezincification—localized dezincification as opposed to uniform or layer-type. See also **dezincification**.

plume—air side discharge from evaporative cooling towers. The condensed water that evaporated from the cooling process and hence is free of chemicals and minerals. Thus it differs from drift. Also a localized concentration of contaminants in groundwater that gradually spreads from the source.

PMB—*abbreviation for* **plastic media blasting**.

pOH—negative logarithm of the hydroxide ion concentration, $-\log(OH^-)$. pOH = 14 – pH. See also **pH**.

point defects—vacancies caused by the absence of one or more atoms in a crystal, impurity atoms of different sizes, and interstitial atoms.

poise—unit of coefficient of viscosity and defined as the tangential force per unit area (dynes/cm^2) required to maintain unit difference in velocity (1 cm/sec) between two parallel planes separated by 1 cm of fluid.

poison—any substance that causes injury, illness, or death by chemical means. Also any substance that prevents the activity of a catalyst.

polar compound (molecule)—compound that is either ionic or that has molecules with a large permanent dipole moment, e.g., water. *Opposite of* **nonpolar compound**.

polar covalent bond—bond in which electrons are shared unequally; one with a separation of charge.

polarization—retardation of an electrochemical reaction (corrosion) caused by various physical and chemical factors. It is the shift in electrode potential resulting from the effects of current flow. Polarization can be of two different types, activation polarization and concentration polarization. Activation polarization refers to an electrochemical process controlled by the reaction sequence at the metal-electrolyte interface, while concentration polarization refers to an electrochemical process controlled by the diffusion in the electrolyte. The three main types of polarization effects are termed electronic, atomic, and orientation polarization. Polarization also refers to the barrier produced at cathodic sites of a metal corroding in an aqueous system from the resulting buildup of hydroxyl ions (OH^-), hydrogen gas (H_2), or both. This slows the movement of oxygen gas (O_2) or hydrogen ions (H^+) to the cathode, and thus the barrier becomes a corrosion inhibitor. Removal or disruption of this barrier, usually by lowering water pH or increasing the water velocity into the turbulent flow region, causes a resumption of corrosion. This latter effect is called depolarization.

polarization admittance—reciprocal of polarization resistance.

polarization curve—plot of current density versus electrode potential for a specific electrode-electrolyte combination.

polarization resistance—slope at the corrosion potential of a potential-current density curve.

polarize—to develop a barrier on the anodic or cathodic surface disrupting the corrosion process.

polarized light—light that exhibits different properties in different directions at right angles to the line of propagation.

polar solvent—liquid solvent that has a dipole moment and consequently a high dielectric constant (e.g., water and liquid ammonia) making it capable of dissolving ionic compounds or covalent compounds that ionize.

polishing—abrading operation to improve the surface condition or remove unwanted protrusions from a metal surface. The operation usually follows grinding and precedes buffing.

pollutant—contaminant at a concentration high enough to endanger the aquatic environment or the public health.

pollution—undesirable change in the physical, chemical, or biological characteristics of the natural environment brought on by man's activities.

polyacrylates—organic polymers containing carboxylic acid groups and their derivatives, e.g., acrylic acid (CH_2=CHCOOH), which are used as dispersants of scale and suspended matter in cooling waters.

polyalcohol—organic compounds containing more than one hydroxyl (–OH) group, e.g., ethylene glycol, glycerol. Sometimes called polyhydric alcohols or polyols.

polyamide—condensation polymer produced by the interaction of an amino group ($-NH_2$) of one molecule and a carboxylic acid group ($-COOH$) of another to give a chain. Forms of the polymer are also known as nylon.

polyatomic ion—ion containing more than one atom.

polycarbonate plastics—polyester plastics based on polymers in which the repeating structural units are essentially of the carbonate type CO_3^{2-}.

polycyclic compounds—compounds containing two or more fused ring structures within their molecules.

polyelectrode (mixed electrode)—metallic surface on which there is a simultaneous occurrence of two electrochemical reactions (anodic and cathodic) at a metal electrolyte interface with the result being corrosion.

polyelectrolyte—water-soluble macromolecular organic polymeric material with incorporated ionic constituents and having ion-exchange sites on its skeleton. Used for clarifying water whether they function as primary coagulants or flocculants. Classified as anionic, cationic, or nonionic.

polyester—polymer in which the repeated structural unit in the chain is of the ester type, i.e., condensation product of the interaction of a polyhydric alcohol and polybasic acid.

polyethylene (PE)—polymer prepared by the polymerization of ethylene (C_2H_4) as the sole monomer.

polymaleic anhydride—organic polymer prepared by the polymerization of maleic acid (HCOOHC=CHCOOH) and used for scale and suspended matter control in cooling waters. It forms the polymeric acid structure when solubilized in the water.

polymer—macromolecular material formed by the chemical combination of monomers having either the same or different composition.

polymer corrosion—degradation (corrosion) of polymeric materials by physicochemical rather than electrochemical processes. The degradation is usually bond rupturing due to oxidation, excessive heat, or intense radiation.

polymeric polyol—polymer molecule with multiple pendant hydroxyl groups.

polymerization—chemical reaction in which small molecules combine to form large molecules.

polymethacrylates—organic polymers similar to the polyacrylates in structure and application but with methyl groupings affixed to the carbon backbone, e.g., $CH_2=C(CH_3)COOCH_3$.

polymorphism—ability to exist in two or more crystalline forms.

polyolefin—polymer prepared by the polymerization of an olefin, e.g., $RCH=CH_2$, as the sole monomer.

polyphosphate—molecularly dehydrated orthophosphate, e.g., $NaPO_3$ and its condensed products $(NaPO_3)_x$. Also refers to commercially available condensed phosphates extending from simple pyrophosphate ($P_2O_7^{4-}$) to linear, long-chain polymeric structures of about 21 repeating oxygen/phosphorus units. Used by water utilities for corrosion and scale control, metal ion stabilization, and dispersion.

polypropylene (PP)—polymer prepared by the polymerization of propylene ($CH_3CH=CH_2$) as the sole monomer.

polyprotic acid—acid that has the capacity to donate more than one hydrogen ion (H^+).

polystyrene (PS)—polymer prepared by the polymerization of styrene ($C_6H_5CH=CH_2$) as the sole monomer.

polysulfide—elastomer whose chains contain the unit –C–S–S– and which has very good resistance to ozone and ultraviolet radiation.

polyterephthalate (PET)—thermoplastic polyester in which the terephthalate (from 1,2-benzenedicarboxylic acid) group is a repeated structural unit in the chain.

polytetrafluoroethylene (PTFE)—fluoropolymer having a service temperature to 245 to 260 °C (475 to 500 °F) and, except for molten alkali metals and free fluorine, is immune to most corrosive environments.

polythionic acid cracking—stress-corrosion cracking of austenitic stainless steels used in petroleum refining and petrochemical operations by polythionic acids (oxo acids of sulfur with the general formula $HO \cdot SO_2 \cdot S_n \cdot OH$ where n is 0 to 4).

polythionic acids—*see* **polythionic acid cracking**.

polyurethane—polymer containing the urethane group, i.e., –NH·CO·O–, and prepared by the reaction of an organic diisocyanate with compounds containing hydroxyl groups. *See also* **urethane resins**.

polyvinyl acetate (PVAc)—polymer prepared by the polymerization of vinyl acetate ($CH_2=CHOOCCH_3$) as the sole monomer.

polyvinyl alcohol (PVA)—water-soluble polymer prepared by the essentially complete hydrolysis of polyvinyl ester.

polyvinyl chloride (PVC)—polymer prepared by the polymerization of vinyl chloride ($CH_2=CHCl$) as the sole monomer.

polyvinylidene fluoride (PVDF)—crystalline, high molecular weight, partially fluorinated polymer of vinylidenedifluoride having a service temperature of about 135 °C (275 °F), but because it is not fully fluorinated it is subject to chemical attack by hot aromatic solvents, chlorinated hydrocarbons, ethers, methanol, butanol, and ketones above 65 °C (150 °F).

porcelain enamel—vitreous or glassy inorganic coating bonded to metal by fusion at a temperature above about 425 °C (800 °F). Applied primarily to products made of sheet steel, cast iron, or aluminum to improve appearance and protect the metal surface. Classified as either ground coat or cover coat enamels. Ground coat enamels contain oxides that promote adherence of the enamel, and cover coat enamels are applied over ground coats to improve appearance and properties of the coating. *Also called* **vitreous enamel**.

pore—internal cavity in a material that may be exposed by cutting, grinding, or polishing to become a pit or hole.

porosity—degree of integrity or continuity in freedom from pores. Also the ratio of empty space to total volume of a material.

Portland cement—hydraulic cement made by heating a mixture of limestone and clay in a kiln and pulverizing the resulting clinker. It consists essentially of hydraulic calcium silicates.

Portland cement coatings—cementatious coatings of Portland cement applied on and used to protect cast iron or steel water pipes on the water side, soil side, or both.

post-cure—heat or radiation treatment to which a cured coating is subjected to enhance one or more of its properties.

potable water—water satisfactory to use for drinking and culinary purposes.

potash—potassium oxide (K_2O). Also term loosely used to describe either potassium carbonate (K_2CO_3) or potassium hydroxide (KOH).

potassium ferricyanide test—test for cleanliness of ferrous surfaces. Paper impregnated with a solution of potassium ferricyanide and other reagents then dried is laid on the metal surface and has more of the solution poured over it. Clean areas will develop a relatively dark blue image on the test paper, whereas oil-contaminated regions will remain colorless or develop yellowish zones.

potential difference—difference in the values of the electric potentials at two points in an electric field or circuit, i.e., it is the work done in moving unit charge from one point to the other.

potential, electric—symbol V. Energy required to bring unit electric charge from infinity to the point in an electric field at which the potential is being specified. Also the potential energy of a unit charge at any point in an electric circuit measured with respect to a specified reference point in the circuit or to ground. Also the driving influence of an electrochemical reaction. Usually expressed in volts or millivolts. *See also* **electromotive force** (EMF or emf).

potential energy—stored energy that results from the position, condition, or composition of matter.

potential-pH diagram—*see* **Pourbaix diagram**.

potentiodynamic (potentiokinetic)—technique for varying the potential of an electrode in a continuous manner at a preset rate.

potentiometer—instrument for measuring electromotive force by balancing against it an equal and opposite electromotive force across a calibrated resistance carrying a definite current.

potentiometric titration—titration in which the end point is found by measuring the potential on an electrode immersed in the reaction mixture.

potentiostat—instrument that maintains an electrode at a constant potential. Used in anodic protection devices and to compile E (potential) log I (current) curves.

potentiostatic—technique for maintaining a constant electrode potential.

POTW—abbreviation for Publicly Owned Treatment Works.

poultice corrosion—term used in the automotive industry to describe the corrosion of vehicle body parts due to the collection of road salts and debris on ledges and in pockets that are kept moist by weather and washing. It is a special case of crevice corrosion. *Also called* **deposit corrosion** and deposit attack.

Pourbaix diagram—graphical representation showing regions of thermodynamic stability of species in metal-water electrolyte systems.

pour point—lowest temperature at which a liquid is observed to flow when cooled and examined under prescribed conditions. Usually applied to petroleum oils.

powder coating—100% solids coating applied as a dry powder and subsequently formed into a film with heat.

power—rate at which work is done or energy is transferred.

power brush cleaning—use of power-driven wire, abrasive filament, and synthetic and natural fiber-filled rotary brushes to clean, deburr, edge blend, polish, or otherwise surface finish a metal surface.

power factor—cosine of the angle between the voltage applied and the current resulting.

power tool cleaning—use of pneumatic and electric portable power tools to prepare a substrate for coating.

power washers—relatively low-pressure water blast-cleaning equipment.

pozzolan—siliceous or aluminous material which will, in finely divided form and in the presence of moisture, chemically react with calcium hydroxide to form compounds of cementatious properties.

PPA—abbreviation for the Pollution Prevention Act of 1990.

ppm—abbreviation for parts per million parts.

PQL—*abbreviation for* **practical quantitation limit**.

PRA—abbreviation for the Paint Research Association (British).

practical quantitation limit—defined by the EPA as the lowest level that can be reliably achieved within specified limits of precision and accuracy during routine laboratory operating conditions.

preboiler equipment—consists of the feedwater heaters, feed pumps, and feed lines that are usually constructed of copper, copper alloys, carbon steel, and phosphor bronze.

precious metal—one of the relatively scarce and valuable gold, silver, and the platinum group metals. Also sometimes improperly called the noble metals.

precipitate—insoluble reaction product in an aqueous chemical reaction.

precipitation—separation of a new phase from solid or liquid solution. Also all the liquid and solid forms of water that are deposited from the atmosphere, including rain, drizzle, snow, hail, dew, and hoar frost.

precipitation hardened stainless steels—chromium-nickel stainless steels hardened by an aging treatment at a moderately elevated temperature. Grades may have austenitic, semiaustenitic, or martensitic crystal structures.

precipitation hardening—hardening of a metallic alloy by the precipitation of a constituent from a supersaturated solid solution.

precipitation inhibitor—corrosion inhibitor that acts by forming slightly soluble or insoluble products on the metal surface to protect it from aqueous corrosion, e.g., zinc and calcium compounds. *See also* **adsorption inhibitor**, **corrosion inhibitor**, and **passivation inhibitor**.

precipitation softening—process for removing water hardness, alkalinity, silica, and other contaminants from system water by using lime or lime and soda ash or lime and sodium aluminate to form insoluble compounds that are removed by sedimentation or filtration.

precision—*see under* **statistical terms**.

precleaning—pretreatment of new equipment of cooling water systems to remove grease, oil, mill scale, corrosion products or dirt to provide clean surfaces for promotion of a uniform protective film by corrosion inhibitors. Usually accomplished by circulating a warm (about 60 °C, or 140 °F) cleaning solution of a polyphosphate, surfactant, and antifoam at pH from 5.5 to 7.0 for about 24 hours then blowing down to remove the precleaning chemicals.

precoated steel—steel prepared for subsequent painting by being coated with zinc or its alloys and usually phosphated for improved paint adhesion and resistance to underfilm attack.

precure—controlled state of partial cure before the final cure.

precursor—compound leading to another compound in a series of chemical reactions.

Preece test—acceptance test for coating uniformity for galvanized steel articles by immersion of the article at room temperature in a copper-sulfate solution. Metallic copper is deposited on discontinuities in the coating or where thinly coated areas have been penetrated by the test solution.

prefabrication primer—quick-drying coating applied as a thin film to a metal after cleaning to give temporary protection.

prefilming—refers to the use of special corrosion inhibitors, or their increased concentrations, to rapidly form a uniform impervious film on metal surfaces to stifle the onset of the corrosion reaction on clean metals. Commonly used to protect metals subject to aqueous immersion and that ordinarily do not rapidly form protective films. Subsequently, other inhibitors, or lower inhibitor treatment concentrations, are used to maintain the established protective film.

premix—admixture of several ingredients to be incorporated into a formulation or process as a group as opposed to individually.

preoperational deposit—black coating of magnetic iron oxide (Fe_3O_4) on the surface of new steel produced during metal fabrication.

prepolymer—polymer of degree of polymerization between that of the monomer and the final polymer.

preservative—chemical substance that, when suitably applied to wood, makes it resistant to attack by fungi, insects, marine borers, or weather conditions. Also applies to chemical additives to various formulations to prohibit or retard microbial growth. Also any material additive used to prevent a substance from fermenting through bacterial action.

pressure—force or load per unit area.

pressure-treated wood—wood treated by applying pressure to force a preservative into it.

pretreatment—chemical treatment of unpainted metal surfaces before painting. Also the cleaning and passivating, with a chemical corrosion inhibitor, of mild steel heat exchangers. Also the initial utilization of corrosion inhibitors in cooling waters at two or three times their normal dosage over the first several weeks of use to significantly improve corrosion control and, later, to reestablish a protective film upon system upsets, pH excursions, corrosive contamination, and prolonged low inhibitor control levels.

preventive maintenance painting—spot repair painting to forestall rusting.

PRI—abbreviation for the Paint Research Institute.

primary cell—voltaic cell in which the chemical reaction producing the emf is not satisfactorily reversible, and the cell cannot therefore be recharged by the application of a charging current.

primary current distribution—current distribution in an electrolytic cell free of polarization.

primary passive potential (passivation potential)—the potential corresponding to the maximum active current density (critical anodic current) of an electrode that exhibits active-passive corrosion behavior.

primary pollutant—pollutant emitted directly from a polluting source. *See also* **secondary pollutant**.

primer—first of two or more coats of a paint, varnish, or lacquer system. *Also called* **basecoat** and **undercoat**.

priming—application of a primer.

printing ink—any fluid or viscous composition of materials used in printing, impressing, stamping, or otherwise transferring on a surface for purposes of information, identification, or decoration.

process additives—chemicals that are injected continuously or intermittently into certain petroleum process streams and includes antifoulants, corrosion inhibitors, neutralizers, and emulsion breakers.

process water—industrially utilized water. Largest use for process water is for cooling the plant and its process equipment. Process water also includes that water used to cool reaction vessels, to generate steam for process heat or electricity, and for process applications such as hydraulic conveying and classification, for washing and equipment cleanup, fire protection, as part of finished products, cutting fluids, etc.

producer gas (air gas)—mixture of carbon monoxide and nitrogen made by passing air over very hot carbon. Usually some steam is added to the air and the mixture may contain some hydrogen. Used as a fuel in some industrial processes.

product—end result of a chemical reaction.

profile—surface contour as viewed from an edge.

prohesion test—cyclic accelerated corrosion test for coated test panels utilizing an electrolyte of 0.4% ammonium sulfate and 0.05% sodium chloride sprayed on at room temperature then dried off at elevated temperature with rapid cycling between spray and dry.

promoter—substance added to a catalyst to increase its activity.

property—physical or chemical characteristic of a sample of matter.

proprietary—being owned by a private individual or corporation under a trademark or patent. Also refers to a product or process available under a brand name.

protection potential—potential below which crevice corrosion and pitting corrosion will not occur. Above this potential, crevice corrosion will occur, but pits will not initiate, although if present they will grow. *See also* **crevice corrosion** and **pitting**.

protective coating—any material composed essentially of synthetic resins or inorganic silicate polymers that provides a continuous coating resistant to industrial or marine environments and prevents serious breakdown or contamination of the underlying structure in spite of abrasion, holidays, or imperfections in the coating. It outperforms conventional paint coatings in these qualities.

protective colloid—material such as a gum, starch, protein, polyacrylate, and cellulose used for protecting charged colloidal particles in aqueous media against flocculation.

protective life—interval of time during which a paint system protects the substrate from deterioration.

protective potential—term used in cathodic protection to define the minimum potential required to suppress corrosion.

protective potential range—range of corrosion potential values in which an acceptable corrosion resistance is achieved for a particular purpose.

protein—any of a group of complex nitrogenous organic compounds of high molecular weight that contain amino acids as their basic structural units and that occur in all living matter.

proton—elementary particle that is stable, bears a positive charge equal in magnitude to that of the electron, and has a mass 1836.12 times that of the electron. The proton occurs in all atomic nuclei.

proton number—*see* **atomic number**.

pseudoplastic fluid—non-Newtonian fluid that does not possess thixotropy. Pressure starts the fluid flow, and the viscosity decreases instantaneously with increasing rate of shear until, at a given point, the viscosity becomes constant.

psychrometer—test instrument used to determine humidity and dew point.

psychrometer, sling—*see* **sling psychrometer**.

PTFE—*abbreviation for* **polytetrafluoroethylene** resin.

Puckorius practical scale index (PSI)—used to predict the calcium carbonate scaling tendency of cooling water. The PSI index value is equal to twice the pH of calcium carbonate saturation minus the equilibrium pH value. The latter is equal to 1.465 log total alkalinity plus 4.54. PSI indices less than 6.0 are scaling. *See also* **Langelier saturation index (calcium carbonate saturation index)** and **Ryzner stability index**.

pulp—fibrous matter derived from cellulosic fiber-containing materials. Also a soft, moist, shapeless mass of matter.

pulpwood—wood cut or prepared primarily for the production of wood pulp.

pulsed cavitation test—procedure using a vibratory cavitation device in which the cavitation is generated intermittently with alternating vibratory and quiescent periods of specified duration.

pumice—porous volcanic rock, usually granitic silicates, that is light and full of cavities from expanding gases that were liberated from solution in the lava while it solidified. The powdered form is used as an abrasive and for polishing.

punching—shearing of holes in sheet metal with a punch and die.

pure substances—matter that has a definite, or fixed, composition, i.e., elements or compounds. Contrast with mixtures whose composition may vary.

putty—doughlike material consisting of pigment and vehicle used to seal glass in frames and cover imperfections in wood and metal surfaces.

PVA—*abbreviation for* **polyvinyl alcohol** and, less often, **polyvinyl acetate**.

PVC—*abbreviation for* **polyvinyl chloride** resin and **pigment volume concentration**.

PVD—*abbreviation for* **physical vapor deposition** covering a broad class of vacuum coating processes.

PVDF—*abbreviation for* **polyvinylidene fluoride**.

PWR—*abbreviation for* **nuclear pressurized-water reactor**. *See also* **boiling-water reactor** and **light-water reactor**.

pyrite—mineral form of iron(II) sulfide (FeS_2).

pyrolysis—chemical decomposition occurring as a result of high temperature but without oxidation. Charring.

pyrometer—instrument for measuring temperatures beyond the upper limit of the usual liquid thermometer and that usually operates either on the basis of differential expansion of dissimilar metal strips, changes of resistance, current flowing through two joined pieces of metal, heat radiated from a hot body, or the intensity of light emitted from a hot body.

pyrometric cones—series of conical materials used to indicate the temperature inside a furnace or kiln. Most commonly used in ceramic materials processing. The cones are made from different mixtures of clay, limestone, feldspar, and other minerals and each cone, depending on its composition, softens and droops at a different temperature.

pyrophosphate—the ion ($P_2O_7^{4-}$).

QA—abbreviation for quality assurance.

QC—*abbreviation for* **quality control**.

QCT—*See* **Cleveland Condensing Humidity Cabinet**.

qualification procedure—defined sequence of actions or functions to establish or verify a specified level of quality or competence.

qualification test—testing performed on a product to determine whether or not the product conforms to the requirements of an applicable specification.

qualitative analysis—analysis in which some or all of the components of a sample are identified.

quality control—system whereby a manufacturer ensures that materials, methods, workmanship, and the final product meet the requirements of a given standard.

quantitative analysis—analysis in which the amount of some or all of the components of a sample are determined.

quartz—most abundant mineral; crystalline silicon dioxide (SiO_2).

quaternary ammonium—the organic group $-N(CH_3)^{4+}$. Some biocides utilize quaternary ammonium salts.

quebracho—drilling fluid additive for thinning or dispersing to control fluid viscosity and thixotropy. A crystalline extract of the quebracho tree consisting essentially of tannic acid.

quench—to rapidly cool a metal or alloy from an elevated temperature by means of liquid immersion in water, salt solution, oil, or by forced-air cooling to prevent it from reaching a stable condition and in so doing imparts special properties to the metal or alloy.

quench annealing—method used to minimize intergranular corrosion of austenitic stainless steels by their high-temperature solution heat treatment at 1060 to 1120 °C (1950 to 2050 °F) followed by water quenching. Method dissolves chromium carbide and a more homogeneous alloy develops. Also called solution quenching.

quenching and tempering (hardening and tempering)—quenching of, usually, steel into water or oil transforming it to martensite and increasing internal stresses which are relieved by heating (tempering) up to 100 to 200 °C (212 to 392 °F) to diffuse carbon and precipitate small carbides to improve the strength and ductility of the material.

quick-hardening lime—hydraulic lime. *See also* **quicklime**.

quicklime—calcium oxide (CaO) made by calcining limestone ($CaCO_3$). Capable of slaking with water to form calcium hydroxide ($Ca(OH)_2$) for use as an inexpensive alkali to neutralize the acidity in certain soils or for other applications. *Also called* **burnt lime** and **lime**. *See also* **calcium oxide**.

quinone—any organic compound containing C=O groups in its unsaturated ring.

R

radiant energy—energy transmitted as electromagnetic waves.

radiation—emission and propagation of energy through space or through a material medium in the form of waves.

radiation corrosion—corrosion damage to metals resulting from their exposure to intense radiation, i.e., neutrons or other energetic particles, causing lattice changes resembling those produced by severe cold working, e.g., lattice vacancies, interstitial atoms, and dislocations, such as to increase the diffusion rate of specific impurities or alloyed components.

radiation curable coatings—paint systems that are cured mainly by ultraviolet radiation (UV), infrared radiation (IR), electron beam radiation (EB), radio frequency radiation (RF), or microwaves.

radiation damage—general term for the alteration of properties of a material caused by exposure to ionizing radiation such as X-rays, gamma rays, neutrons, heavy-particle radiation, or fission fragments.

radiator cleaners—aqueous compositions used separately in automotive and truck cooling systems to dissolve or dislodge and hold in suspension deposits that had formed and collected on the walls of the engine water jacket, radiator, thermostat, heater core, radiator hose, etc.

radical—group of atoms, either in a compound or existing alone, that behave as a single unit.

radioactive element—element that spontaneously emits various matter/energy forms at a measurable rate.

radioactivity—spontaneous nuclear disintegration with emission of corpuscular or electromagnetic radiation, or both.

radio frequency (RF)—electromagnetic radiation used in radio transmission that encompasses a range of frequencies between about 3 kilohertz and 300 gigahertz.

radio frequency interference (RFI)—any electromagnetic radiation interfering with the transmission of radio frequency. *See also* **radio frequency**.

radiographic inspection—use of X-rays or nuclear radiation to detect discontinuities in material. Also used to indicate distribution or depth of corrosion attack.

radiography—nondestructive test method employing penetrating X-rays, gamma rays, or other forms of radiation to provide a permanent visible film record of internal conditions of a structure.

radioisotopes—radionuclides having the same atomic number. Isotope of an element that is radioactive.

radio-opaque—medium that is opaque to X-rays and gamma rays. Opposite of radio-transparent.

rag—debris that accumulates at an oil-water interface.

rain density—mass of liquid per unit volume of mixture in an actual or simulated rainfield.

rainfall—*see* **precipitation**.

Raman spectroscopy—direct corrosion monitoring method that reflects an infrared beam off a surface and records the change in frequency of the beam as the Raman spectrum. Comparison with known Raman spectra is used to identify constituents of the corrosion film.

Raman spectrum—spectrum resulting from the alteration in frequency and random alteration in the phase of monochromatic light scattered on passing through a transparent substance.

random—having no specific pattern or objective. In statistics, designating a phenomenon that does not produce the same outcome or consequences every time it occurs under identical circumstances.

random error—*see under* **statistical terms**.

range—*see under* **statistical terms**.

Rankine cycle—successive stages in heat content and temperature as water is converted to steam, expands through a prime mover, condenses, and returns to the boiler.

Rankine temperature—temperature scale often used in engineering practice. The Rankine degree is the same as that of the Fahrenheit degree, but the zero point of the Rankine scale is the same as that of the Kelvin scale except in Rankine degrees or 491.69 R.

Raoult's law—molar weights of nonvolatile nonelectrolytes when dissolved in a definite weight of a given solvent under the same conditions lower the solvent's freezing point, elevate its boiling point, and reduce its vapor pressure equally for all such solutes. Also states that the partial vapor pressure of a solvent is proportional to its mole fraction.

rare earth element—one of 15 chemically similar metals with atomic numbers 57 (lanthanum) through 71 (lutetium). *Also known as* **lanthanides**.

rare gases—*see* **noble gases**.

rate-determining step—slowest step in a chemical reaction that involves a number of steps. The rate of this step determines the overall rate of the reaction.

raw water—untreated surface and ground water containing varying amounts of dissolved and suspended matter depending on its source.

rayon—textile made from cellulose by either viscose or acetate processes.

RCRA—*abbreviation for* (U.S.) **Resource Conservation and Recovery Act**.

reactant—substance that is initially present in a chemical reaction.

reaction—any change in chemical composition accompanied by a change in enthalpy.

reactive diluent—volatile compound that, by itself, evaporates on exposure to the atmosphere, but when mixed with certain resins may react to form part of the solids of the coating and, once reacted, is no longer volatile and cannot evaporate. Examples of a reactive diluent include the styrene monomer used in some polyester gel coatings, and the low molecular weight resins used in UV curable and electron beam coatings.

reactive metal—metal that readily combines with oxygen.

reactive pigments—pigments that react with the vehicle, e.g., the formation of zinc and lead soaps, by the interaction of those basis pigments with drying oils.

reagent—substance reacting with another substance.

real gas—gas that does not behave exactly in the manner predicted by the ideal gas laws.

rebar corrosion—corrosion of steel reinforcing bars in concrete.

recarbonation—injection of carbon dioxide into water to decrease its pH after lime softening.

recharge zone—area of land that serves as the primary source of water to replenish an aquifer.

rectification—conversion of alternating into direct current.

rectifier—device that is provided with alternating current electrical power and converts this to a lower voltage direct current by means of a stepdown transformer and rectifying device.

recovery furnace—furnace for burning black liquor from the kraft pulping process to recover the smelt. *See also* **smelt**.

red brass—copper alloy nominally consisting of 85% copper and 15% zinc; this alloy is relatively immune to dezincification, but more susceptible to impingement attack than yellow brass.

red lead—dilead (II) lead (IV) oxide, Pb_2PbO_4 (or Pb_3O_4); a red amorphous powder used as a corrosion-inhibiting pigment in paints.

redox—contraction term for reduction and oxidation.

reduced—undergoing a gain of one or more electrons. *Contrast with* **oxidized**.

red oxide primer—primer paint based on natural red iron oxide (essentially Fe_2O_3) as the major pigment component.

redox potential—reduction-oxidation potential measured against a standard electrode.

reducer—material that lowers paint viscosity, but is not necessarily a solvent for the particular film former. A thinner.

reducing agent—compound or element that causes reduction of another substance and that itself is oxidized. Also known as a reductant.

reducing atmosphere corrosion—form of gas-phase corrosion that may occur in fossil fuel power plants. Usually the result of direct reaction of the water-wall tubes with a substoichiometric gaseous environment containing sulfur or with deposited, partially combusted char containing iron pyrites, which gives rise to a very localized corrosive environment. The reducing conditions tend to lower the melting temperature of any deposited slag which, in turn, may dissolve normal oxide scale on the tubes, or the H_2S may preferentially form less protective iron sulfide rather than iron oxides.

reduction—gain of electrons by a constituent of a chemical reaction.

reference electrode—electrode against which the electrical potential of a specimen is measured, e.g., hydrogen electrode and saturated calomel electrode.

reference material—material of defined chemical composition or chemical or physical properties.

reference standard—standard used to calibrate testing apparatus and methods.

refining—process of purifying a substance or extracting substances from mixtures.

reforming—petroleum-refining process for the conversion of straight-chain alkanes into branched-chain alkanes by cracking or by catalytic reaction.

refractive index—ratio of the velocity of radiation in the first of two media to its velocity in the second as it passes from one into the other.

refractometer—any of various instruments used to measure the refractive index of a substance or medium.

refractories—nonmetallic materials having chemical and physical properties making them applicable for structures or as components of systems that are exposed to environments above about 538 °C (1000 °F).

refractory brick—nonmetallic refractory bricklike materials of construction for use in high-temperature applications and usually further classified as high-alumina bricks, silica bricks, magnesite bricks, chrome bricks, insulating firebrick, high-burned kaolin refractory bricks, fused mullite bricks, fused alumina bricks, fused magnesite bricks, silicon carbide bricks, and zirconia and zircon bricks.

refractory metals—metals or their alloys having a very high melting point as compared with iron and steel. Also those metals having boiling points above 4000 °C (7232 °F). Includes such metals as niobium, molybdenum, tantalum, tungsten, hafnium, and zirconium.

refrigeration—to cool or chill a substance or chamber below the temperature of the external environment.

refrigerator—apparatus for reducing and maintaining the temperature of a chamber below the temperature of the external environment. It can be thought of as a heat engine run backward.

regeneration—part of the operating cycle of an ion-exchange process in which a specific chemical solution is passed through the ion exchange bed to prepare it for use.

regenerative heating—in utility stations, a scheme for reducing heat losses to the main condenser in the cycle by using steam extracted from the turbine to heat feedwater. In engineering designs, it is the use of a heat exchanger to preheat the feed to a process by extracting heat from the product.

Registry of Toxic Effects of Chemical Substances (RTECS)—accession number used by the Registry of Toxic Effects of Chemical Substances produced by the National Institute for Occupational Safety and Health, U.S. Department of Health and Human Services. It is useful for locating published toxicity data.

regenerator—cyclic heat exchanger that alternately receives heat from gaseous combustion products and transfers heat to air or gas before combustion.

reheater—heat exchanger located in a furnace to increase the temperature of steam extracted from a turbine for reinjection.

reinforced plastic—plastic with high-strength fillers embedded in the composition resulting in an improvement in some mechanical properties. Also a plastic with strength properties greatly superior to those of the base resin due to the presence of reinforcements in the composition.

relative humidity—ratio, expressed as a percentage, of the amount of moisture in the air compared with what it could hold if saturated at the temperature involved.

relative humidity test, 100%—see ASTM D 2247 for details.

relative permittivity—*see* **dielectric constant**.

relay—electrical or electronic device in which a variation in the current in one circuit controls the current in a second circuit.

reliability—probability of performing without failure a specified function under specified conditions and for specified time.

remediation—actions taken to reduce contamination or the effects of contamination in groundwater or the ground.

repeatability—*see under* **statistical terms**.

replacement reaction—reaction in which an element combines with a compound and displaces one of the components of the compound.

replica—reproduction of a surface of a material. Also a close, but not exact, copy.

replicate—one of two or more runs with the same test conditions. Also to repeat a run so as to produce a replicate.

reproducibility—*see under* **statistical terms**.

rerusting—rusting of previously cleaned, moist ferrous surfaces. A condition that is promoted by mildly acidic residues. It is minimized by rapid dry-off, prior rinsing with aqueous alkaline or dilute sodium nitrite solutions, or by the application of a water displacing rust preventative. *Also called* **rust-back**.

reserve alkalinity—amount of alkaline substances present in a product or solution. Usually determined by titrating 10 milliliters of the sample with $0.100\ N$ hydrochloric acid

(HCl) to a pH of 5.5 and reporting the number of milliliters, to the nearest 0.1 milliliter, of acid required.

residual chlorine—total amount of free and combined chlorine remaining in water after the chlorine demand has been satisfied.

residual stress—stresses that remain within a body as a result of plastic deformation.

residue—any undesirable material remaining on a substance after any process step.

resin—solid or pseudosolid organic material, usually of high molecular weight, that tends to flow when stressed; usually has a softening or melting range and fractures conchoidally. Also used to describe the organic materials of natural or synthetic origin that are used as film formers in coating technology. For general purposes, the terms resin, polymer, and plastic can be used interchangeably.

resin emulsion paint—water-based paint consisting of a water emulsion of an oil-modified alkyd or other resin that when dry leaves a tough film of the resin.

resin, natural—solid organic substance originating in the secretion of certain plants or insects; being thermoplastic, flammable, breaking with a conchoidal fracture, and soluble in organic solvents, but not in water.

resin, synthetic—term applied to a member of a heterogeneous group of organic compounds synthetically produced from simpler compounds by condensation or polymerization reactions.

resistance—ratio of the potential difference applied to a specimen to the current passed through by the applied potential. The unit of resistance measurement is the ohm, which is the reciprocal of conductance.

resistivity—electrical resistance of a body of unit length and cross-sectional area or weight.

resistor—component in an electrical circuit that is present because of its electrical resistance.

resolution—breaking of an emulsion into its separate components.

Resource Conservation and Recycling Act of 1980 (RCRA)—pertains to solid waste disposal, storage, and handling. Identifies wastes subject to regulation, defines terms, establishes management requirements for hazardous waste, and specifies identification and handling procedures.

respiration—metabolic process by which an organism assimilates oxygen and releases carbon dioxide and other products of oxidation.

rest electrode potential—measured potential of an electrode when no net current is flowing across the metal-solution interface. *Also called* **corrosion potential**, freely corroding potential, and **open-circuit potential**.

retarder—material that slows the rate at which chemical reactions would otherwise occur. Sometimes used as a synonym for inhibitor.

reticulation—surface defect of netlike appearance.

reverse osmosis—process that reverses, by the application of pressure, the flow of water in the natural process of osmosis so that it passes from the more concentrated to the more dilute solution. The process requires a pressure of about 25 atmospheres, which makes the process difficult to apply on a large scale for producing fresh water from seawater.

reversible potential—*see* **equilibrium potential**.

reversible reaction—reaction in which the products can combine to produce the reactants.

reversion—return of molecularly dehydrated phosphate (polyphosphate) to its hydrated origin (orthophosphate).

Reynolds number—dimensionless number used in fluid dynamics to determine the type of flow of a fluid through a pipe. It is the ratio vpl/n where v is the flow velocity, p is the fluid density, l is a characteristic linear dimension, such as the diameter of the pipe, and n is the fluid viscosity. Laminar flow occurs if the Reynolds number is less than about 2000, while turbulent flow is established at above 3000.

RFI—*abbreviation for* **radio frequency interference**.

rheology—study of the deformation and flow of matter.

rheostat—variable resistor, the value of which can be changed without interrupting the current flow.

Riddick's corrosion index—applies to soft waters to determine their corrosivity. Its determination utilizes a number of water characteristics including $CaCO_3$ solubility, dissolved oxygen, chloride ion concentration, noncarbonate hardness, and silica concentration.

rigid plastic—generally describes plastics having a modulus of elasticity, either in flexure or in tension, greater than 100,000 psi at 23 °C (73 °F) and 50% relative humidity.

rigidsol—a plastisol having a high elastic modulus, usually produced with a cross-linking plasticizer.

ring—closed chain of atoms in a molecule.

Ringlemann test—method used in air pollution evaluation that compares the opacity of a stack plume to a standard set of disks representing increasing degrees of discoloration.

ringworm corrosion—localized corrosion frequently observed in oil-well tubing where a circumferential attack is observed near a region of metal "upset."

rinse—treatment with water or other solvent for the purpose of removing dirt or other unwanted contaminant from a material.

rising damp—upward-moving moisture in a wall or other structure standing in water or in wet soil.

risk—probability or likelihood an adverse effect will occur.

RMS—*abbreviation for* **root-mean-square value**.

R&O—*abbreviation for* **rust- and oxidation-inhibited**.

road salts—*see* **deicing salts**.

rock—aggregate of mineral particles that makes up part of the earth's crust. Rock may be consolidated or unconsolidated, e.g., sand, gravel, mud, shells, coral, clay.

rock salt—common salt. Crystalline sodium chloride (NaCl) occurring in large solid masses or as the subdivided product from these. *See also* **halite**.

rodding—abrasive cleaning method that passes rods, rubber plugs, or nylon brushes through heat-exchanger tubes to physically remove deposits.

rolling friction—friction between a rolling wheel and the plane surface on which it is rotating.

room temperature—loose term for temperature in the range of 20 to 30 °C (68 to 85 °F).

root-mean-square value (RMS)—statistical term describing a typical value of a number of values of a quantity equal to the square root of the sum of the squares of the values divided by the number of values.

rosin—natural resin obtained as a residue in the distillation of crude turpentine from the sap of the pine tree.

rosin acids—monocarboxylic acids with the empirical formula $C_{19}H_{29}COOH$, which are found in wood, gum, and tall oil rosins.

rosin amine salts—used as algicides in water systems. *See also* **rosin acids**.

rosin oil—oily portion of the condensate obtained when rosin is dry-distilled.

rot—referring to wood, it is decomposition by fungi and other microorganisms resulting in reduced strength, density, and hardness.

rotary converter—electric motor coupled to a generator device for converting direct current to alternating current or one voltage to another.

rotary peening—surface peen and preparation technique utilizing a portable power-tool system, three-dimensional nonwoven surface conditioning products, and an impacting and peening flap wheel.

rotor—rotating part of an electric motor, electric generator, or turbine. *See also* **stator**.

rottenstone—brown, amorphous, siliceous limestone, similar to pumice, but softer in texture. Used as an abrasive. *See also* **tripoli**.

rouge—reddish powder, chiefly, ferric oxide (Fe_2O_3) used to polish metals.

round-robin test—practice of planning, conducting, analyzing, and interpreting the results of interlaboratory tests on the performance results, or chemical and physical properties of a test chemical, formulation, or material.

RSI—*abbreviation for* **Ryzner stability index**.

RTECS—*abbreviation for* **Registry of Toxic Effects of Chemical Substances**.

rubber—elastic substance derived from various plants; essentially a polymer of isoprene (CH_2=$C(CH_3)CH$=CH_2). The term rubber is also frequently applied to both natural and synthetic elastic substances.

Rule 66—Los Angeles Air Pollution Control District rule that restricts the amount of photochemically reactive smog-causing solvent vapors that can be evaporated into the atmosphere.

rural environment—describes a natural atmospheric exposure that is virtually unpolluted by smoke and sulfur gases, and which is sufficiently inland to be unaffected by salt contaminations or the high humidities of coastal areas.

rust—visible reddish-brown corrosion product consisting of hydrated oxides of iron(III) ($Fe_2O_3 \cdot xH_2O$). It is an electrochemical process that occurs only in the presence of both water and oxygen and is accelerated by the presence of acids or other electrolytes in the water. Applied only to ferrous alloys.

rust- and oxidation-inhibited (R&O)—highly refined industrial lubricating oils formulated for long service in circulating systems, compressors, hydraulic systems, bearing housings, gear cases, etc. Sometimes also called turbine oils.

rust-back (rerusting)—rusting of freshly cleaned ferrous surface on exposure to conditions of high humidity, moisture, contamination, or a corrosive atmosphere.

rust bloom—discoloration indicating the beginning of rusting.

rust converters—treatments to remove surface rust by immersion, spraying, or brushing with aqueous compositions based on phosphoric acid, tannins, or oximes, and that frequently also incorporate organic solvents and polymers. Those based on tannins and oximes react to form organometallic complexes rendering the rusty steel surface suitable for painting without need for abrasive blasting.

rust grade scale—linear, numerical rust grade scale being an exponential function of the area of rust so that slight amounts of rust have the greatest effect on lowering the rust grade. See ASTM D 610 for details.

rusting—corrosion of iron or iron-base alloys to form a reddish-brown surface product which is primarily hydrated ferric oxide ($Fe_2O_3 \cdot xH_2O$).

rusting failure of coated steel—may appear as any of (1) spot rusting at damaged areas, (2) pinhole rusting at minute areas, (3) rust nodules breaking through the coating, and (4) underfilm rusting that eventually causes peeling or flaking of the coating.

rust-inhibitive washes—conversion coating treatment of ferrous surfaces prior to their subsequent painting.

rust prevention test for turbine oils—test measuring the effectiveness of an oil in preventing the rusting of ferrous parts in the presence of water. See ASTM D 665 for details.

rust-preventive compounds—easily removable coatings used to protect iron and steel surfaces against corrosive environments during fabrication, storage, or use. Composed of a basis coating that usually incorporates corrosion inhibitors and other additives. Generally categorized by their basis coating material type as being either petrolatum compositions having a greaselike consistency; oily compositions; hard, dry film compositions; solvent cutback petroleum residual compositions; emulsion compositions; or water-displacing polar compositions. A special category and application for rust-preventive compounds are the "fingerprint removers." *See also* **fingerprint removers** and **temporary coating**.

rustproof—being resistant to rusting.

rustproofing oils—oils used to provide rust protection to iron and steel in storage or transport and that usually contain a soluble metal sulfonate or fatty acid metal soap for corrosion inhibition and moisture barrier additives such as lanolin, microcrystalline wax, or oxidized petrolatum wax.

rust (and scale) removal—most commonly employed methods for rust and scale removal from ferrous mill products, castings, forgings, and fabricated parts include: (a) abrasive blasting (dry or wet), (b) brushing, (c) tumbling (dry or wet), (d) acid pickling, (e) salt bath descaling, (f) alkaline descaling, and (g) acid cleaning.

rust-resistant—term used to describe a material designed to inhibit rusting.

Ryzner stability index (RSI)—means of expressing the degree of saturation of water relative to its likelihood for scale-forming. A qualitative indication to predict the tendency of calcium carbonate in a cooling water to deposit or dissolve. Differs from the Langelier saturation index (also known as the calcium carbonate saturation index) in that it distinguishes between waters of low hardness and high hardness. The Ryzner stability index is the algebraic difference between twice the saturation pH (*see* **Langelier saturation index**) and the actual pH. *See also* **corrosion index**.

S

sacrificial anode—*see* **anode, sacrificial**. *See also* **cathodic protection** and **sacrificial protection**.

sacrificial protection—reduction or prevention of corrosion of a metal in an electrolyte by galvanically coupling it to a more anodic metal. *See also* **anode, sacrificial** and **cathodic protection**.

SACV—abbreviation for small-amplitude cyclic voltammetry.

SAE—abbreviation for the Society of Automotive Engineers.

SAE oil viscosity classification (SAE numbers)—standardized numbers applied to crankcase, transmission, and rear-axle lubricants to indicate their viscosity range.

sal ammoniac—ammonium chloride (NH_4Cl).

saline—solution containing a dissolved salt; usually refers to dissolved sodium chloride (NaCl).

salinometer—instrument for measuring the salinity of a solution. Examples include a hydrometer, which measures solution density, and various apparatus measuring solution electrical conductivity.

salinity—concentration of soluble minerals in water.

saliva—body fluid secreted by the salivary glands that helps lubricate the mouth and esophagus and that contains mucin for lubrication and digestive enzymes.

sal soda—anhydrous sodium carbonate (Na_2CO_3).

salt—compound formed by reaction of an acid with a base in which the hydrogen of the acid has been replaced by a metal or other positive ion. Any substance that yields ions other than hydrogen or hydroxyl ions. Also the common term for sodium chloride (NaCl).

salt bath descaling—oxidizing, electrolytic, or reducing fused salt cleaning processes for removal of oxide heat treat scales, free graphite, lubricant and binder residues, glass-type lubricants, mold and core material, and core washes from metals.

salt bridge—electrical connection made between two half cells using a conductive salt solution.

salt cake—industrial grade sodium sulfate (Na_2SO_4).

salt fog (spray) test—accelerated corrosion test in which specimens are exposed to a fine mist of an aqueous solution of sodium chloride but that may also incorporate other chemicals. See ASTM B 117.

salt splitting—ability of a cation-exchanger to convert a salt solution to acid and an anion-exchanger to convert a salt solution to a caustic solution.

salt spray test—*see* **salt fog (spray) test**.

saltwater muds—oil field drilling fluid containing dissolved salt.

SAM—abbreviation for scanning Auger microprobe.

sampling—obtaining a representative portion of the material of concern.

sand—particles of rock with diameters generally in the range 0.06 to 2.00 millimeters. Most commonly composed of quartz (silicon dioxide, SiO_2).

sand blasting—abrasive blasting with sand, flint, or similar abrasive when propelled by an air blast.

sandpaper—paper coated with an abrasive material such as silica, garnet, silicon carbide, or aluminum oxide and used for smoothing and polishing surfaces.

sand scouring—use of sandblasting equipment to remove light scale from the tube side of heat exchangers. Usually accomplished in combination with backwashing.

sanitary landfill—place of disposal of solid wastes where they are piled into a trench, compacted, and covered over.

sanitary ware—porcelain enameled ware, e.g., sinks, lavatories, and bathtubs.

sanitizer—chemical designed to kill sufficient numbers of bacteria and other microorganisms to avoid problems with disease or excessive spoilage.

sap—moisture in unseasoned wood that also contains nutrients and other chemicals in solution.

saponification—alkaline hydrolysis of fats whereby a soap is formed. Also the process by which paint films are softened and stripped by alkalis and due to the alkaline nature of the cathodic area during underfilm corrosion.

sapwood—wood containing some living cells and forming the initial wood layer under the bark. The layers of wood next to the bark and that are usually lighter in color than the heartwood. More susceptible to decay than heartwood.

SARA—abbreviation for the U.S. Superfund Amendments and Reauthorization Act.

saturant—the substance, solid or liquid, that saturates something else.

saturated—holding the maximum amount of saturant it is capable of holding.

saturated calomel electrode (SCE)—reference electrode composed of mercury coated with mercurous chloride (HgCl) (calomel), and the electrolyte is a saturated aqueous solution of potassium chloride and calomel. Its potential is –0.2415 volt at 25 °C (77 °F).

saturated compound—compound whose elemental valences are satisfied. Also organic compound having no double or triple bonds and are therefore incapable of absorbing substances by addition.

saturated hydrocarbon—hydrocarbons that contain only carbon-carbon single bonds; same as alkanes.

saturated solution—solution in which the maximum amount of solute is dissolved in a solvent for a particular set of conditions.

saturated steam—steam at 100% quality, i.e., no water particles, and at saturated temperature where the steam pressure is equal to the liquid vapor pressure of the water.

saturated zone—region below the ground surface in which all pore spaces are filled with water.

saturation—condition existing when a vapor is in equilibrium with the plane surface of a condensed phase of the same substance (liquid or solid).

saturation index—relationship of calcium carbonate to the pH, alkalinity, and hardness of a water to determine its scale-forming tendency. A positive saturation index describes water oversaturated with respect to $CaCO_3$ and has the characteristic of forming a protective film of $CaCO_3$ on metal surfaces. A zero saturation index water is one that is in equilibrium with dissolving and precipitating $CaCO_3$. A negative saturation index water is undersaturated with respect to $CaCO_3$ and tends to be corrosive toward metals.

SBR—*abbreviation for* **styrene butadiene rubber**.

SCA—*abbreviation for* **supplemental coolant additive**.

scab corrosion—refers to corrosion occurring under a paint film damaged by external means where the paint film remains intact although it does not adhere to the substrate metal.

scale—precipitate or deposit formed on surfaces in contact with water as a result of a physical or chemical change. Also called "fur." Also used to describe the oxide formed on the surface of a metal during or as a result of heating. Also used to describe laminar rust. Scale is also the term used to describe a relatively hard and adherent boiler deposit. Sludge is a softer and less adherent deposit than scale.

scaling—precipitation of dense adherent material on heat-exchange surfaces.

scanning electron microscope (SEM)—microscope that forms a magnified image of the specimen by rastering an electron beam incident on the specimen. A detector records either the secondary or the background scattered electrons emitted from the top side of the specimen near where the electron beam strikes it. The microscope allows for examination of a sample surface at from about 100 to 10,000×. The X-rays emitted also make chemical analysis possible by utilizing energy-dispersive X-ray spectroscopy (EDS).

scarifying—method of preparing concrete surfaces for coating by using a special tool resembling a plant floor sweeper with sharp rotating knives.

scavengers—*see* **oxygen scavengers**.

SCE—*abbreviation for* **saturated calomel electrode**.

scoring—severe form of wear characterized by extensive grooves and scratches in the direction of sliding.

scouring—wet process cleaning by chemical or mechanical means or both.

scrapers—devices used, often in conjunction with pigs, to help remove deposits from pipelines.

scratching—mechanical removal or displacement of material from a surface by the action of abrasive particles or protuberances sliding across the surface.

scratch rusting—rust occurring along a line scratched through a painted surface to the base metal. It is a severe test of the ability of a paint to adhere to surfaces adjacent to a scratch or chip.

scribe creep—measurement taken from the scribe line in a salt spray test to measure the quality or corrosion resistance of a paint film.

scrubber—apparatus used in gas cleaning where the gas is passed through wetted packing or a spray.

SDWA—abbreviation for the U.S. Safe Drinking Water Act.

sealant—adhesive composition formulated to fill gaps and generally having a low bond strength.

sealing—*see* **sealing of anodic coating**.

sealing of anodic coating—process that by absorption, chemical reaction, or other mechanism increases the resistance of an anodic coating to staining and corrosion, improves the durability of the color produced in the coating, or imparts other desirable properties. Sealing is usually accomplished by using any of the following: organic solutions, steam, hot water, nickel acetate, dichromate, etc.

season cracking—cracking of a metal caused by the combined action of corrosion and internal tensile stresses. Usually applied to the stress-corrosion cracking of brass. Commonly occurs when brass is subjected to an applied or residual tension stress while in contact with a trace of ammonia or amine and in the presence of oxygen and moisture.

seawater (ocean water)—natural water of the oceans and seas typically containing about 3.4% sodium chloride plus other cations, anions, living organisms, and dissolved organic nutrients and having a pH about 8. ASTM D 1141 is a specification describing the preparation of substitute ocean water for use in laboratory testing. Seawater produces the most corrosive chloride salt solution obtainable.

seawater muds—saltwater oil field drilling muds for which seawater is used as the fluid phase.

seaweeds—large, multicellular algae living in the sea or in an intertidal zone.

secondary barrier—stable, adherent corrosion-product layer formed on a metal surface that tends to retard further corrosion or make the corrosion rate constant. Commonly observed with aluminum, zinc, manganese, and cadmium.

secondary cell—voltaic cell in which the chemical reaction producing the emf is reversible and therefore the cell can be charged by passing a current through it.

secondary ion mass spectroscopy (SIMS)—analytical technique in which a sample surface is bombarded with selected ions, usually argon, at designated energies, generally less than 1 keV. Secondary ions are sputtered off the sample surface and collected and analyzed with a mass spectrometer.

secondary pollutant—pollutant formed in the atmosphere by chemical changes taking place between primary pollutants and other substances or effects in the air. *See also* **primary pollutant**.

secondary standard (secondary reference standard)—standard calibrated by reference to a primary or other standard.

secondary treatment—second step in most waste-treatment systems in which bacteria consume organic parts of the wastes.

sediment—particulate matter suspended in a fluid and that will settle upon standing.

sedimentation—settling of solid particles in a liquid system to produce a concentrated slurry from a dilute suspension. Also the process by which liquids are clarified through sedimentation.

Seebeck effect—if a circuit consists of two metals, one junction hotter than the other, a current flows in the circuit with the direction of the flow dependent on the metals and the temperature of the junctions.

seed—crystal used to induce other crystals to form from a gas, liquid, or solid.

seepage—infiltration or percolation of water through rock or soil to or from the surface.

segregation—concentration of alloying elements in specific regions in a metal.

selective corrosion—corrosion of certain alloying constituents from an alloy, such as occurs in dezincification, or in an alloy, such as occurs during internal oxidation. *Also called* **selective leaching**.

selective leaching—removal of one element from a solid alloy by corrosion processes. *See also* **dealloying corrosion**, **parting corrosion**, and **selective corrosion**.

selective oxidation—situation where one component of an alloy, usually the most reactive component, is preferentially oxidized (corroded).

selective plating—method of depositing metal from a concentrated electrolyte solution without using immersion tanks. Also known as portable, brush, contact, or spot electroplating. *See also* **brush plating** and **contact plating**.

selectivity—order of preference of an ion-exchange material for each of the ions in the surrounding aqueous environment.

self-curing—coating system that undergoes cure (crosslinking) without the application of heat.

self-priming—use of the same coating for primer and for subsequent coats.

SEM—*abbreviation for* **scanning electron microscope**.

semichemical fluids—*see* **semisynthetic fluids**.

semiconductor—solid with an electrical conductivity intermediate between that of a conductor and an insulator.

semimetal—*see* **metalloid**.

semipermeable membrane—membrane that allows only certain particles to pass through it.

semisynthetic fluids (semichemical fluids)—fluids used in cutting, grinding and other metalworking operations. Consist of mineral, paraffinic, or naphthenic base oil (5 to 30%) emulsified in water plus soluble and emulsified additives to enhance lubricating properties. *See also* **metalworking fluids**. *Also called* **soluble oils**.

sensible heat—heat measurable by temperature alone.

sensitization—heat-treating and welding effect seen particularly with stainless steels, which when heated between 427 and 760 °C (800 and 1400 °F) can form chromium carbide at the grain boundaries such that the chromium is depleted and no longer available to inhibit corrosion, leaving the grain boundaries susceptible to intergranular attack and, further, becoming anodic to surrounding grains.

sequester—to form a stable, water-soluble coordination complex of an ion in solution. *See also* **chelation**.

sequesterant—*alternative term for* **sequestering agent**.

sequestering agent—any compound that in aqueous solution combines with a metallic ion to form a water-soluble combination in which the ion is substantially inactive. *See also* **chelating agent**.

SERS—abbreviation for surface-enhanced Raman spectroscopy.

service run—part of the operating cycle of an ion-exchange process where water is passed through a bed of the ion-exchange material to remove specific ions from the water or to exchange them for an equivalent amount of a specific ion from the bed material.

sesqui-—prefix denoting a ratio of 2:3 in a chemical compound, e.g., a sesquioxide has the formula M_2O_3.

sewage—waste matter carried off by sewers.

sewage sludge—includes, but is not restricted to, domestic septage, scum, or solids removed in primary, secondary, or advanced wastewater-treatment processes and material derived from sewage sludge (EPA definition).

sewer—pipeline for conveying sewage.

shale—thinly stratified and consolidated sedimentary clay with marked cleavage parallel to the bedding. Sometimes contains an organic oil-yielding substance.

shale oil—oil obtained by the destructive distillation of shale tar, which, in turn, is obtained by distilling certain shale deposits.

shear force—force directed parallel to the surface across which it acts. *See also* **shear stress**.

shear strength—maximum shear stress a material is capable of sustaining.

shear stress—force per unit area acting parallel to a plane rather than perpendicularly. A combination of four forces acting over four sides of a plane and producing two equal and opposite couples. *See also* **shear force**.

shelf life—maximum time interval during which a material may be stored and remain in a usable condition. It is usually related to storage conditions.

shellac—alcohol-soluble purified lac, usually prepared in thin orange, yellow, or bleached white flakes or shells by heating and filtering. Lac is a resinous substance secreted by a scale insect and used chiefly in shellac.

shell side—exterior side of heat-exchange tube bundle.

sheltered corrosion—refers to corrosion occurring in locations where moisture condenses or accumulates and does not dry out for long periods of time, e.g., inside surfaces of automobile doors, storage tanks, etc.

sherardizing—process of coating iron or steel with a zinc layer by heating the iron or steel in contact with zinc dust to a temperature slightly below the melting point of the zinc.

shielding gas—stream of inert gas directed at the substrate during thermal spraying or welding operations to envelop the heated area to provide a barrier to the atmosphere in order to minimize oxidation.

shop coat—coating applied in a fabricating shop prior to shipment of the object to a site where the finishing coat is applied. Called shop primer if the coating is a primer only.

shop soils—oil, grease, or cutting fluids as contaminants on the surfaces of parts.

short oil alkyd—alkyd resin containing less than 40% oil in solids.

short oil varnish—varnish made by cooking a relatively small quantity of oil with resin (less than 25 pounds per 100 pounds resin). It is quick-drying and brittle.

Short-Term Exposure Limit—15-minute Time-Weighted Average (TWA) chemical substance concentration exposure that should not be exceeded at any time by a worker during a workday, even if the 8-hour TWA is within the Threshold Limit Value Concentration (TLV-C).

short-term test procedure—corrosion test procedure in which the time to complete an evaluation of a material or system is substantially less than the time for it to fail in service.

shot—small pellets of iron, steel, or any other material that retains its spherical shape when used for peening purposes.

shot blasting—blasting with small spherical objects, such as metallic shot, propelled against a metal surface.

shot peening—method of cold working the surface of a metal by the impingement of a stream of shot at high velocity. It induces compressive stresses in the surface and tends to reduce corrosion fatigue.

SI—abbreviation for the International System of Units whose base units are the meter, kilogram, second, ampere, kelvin (thermodynamic temperature scale), mole, and candela (luminous intensity).

side-stream filtration—use of mechanical filters to remove a portion of the total suspended solids in a recirculating water system by diverting up to about 5% of the circulating flow to the side-stream filter.

siemens—SI unit of electrical conductance equal to the conductance of a circuit having a resistance of 1 ohm. Symbol is S. Formerly called the mho and reciprocal ohm.

sieve—apparatus with square apertures used for separating sizes of material.

sign convention for current density and current—anodic currents and current densities are positive and cathodic currents and current densities are negative.

sign convention for electrode potential—IUPAC sign convention that prescribes the positive direction of electrode potential and implies an increasingly oxidizing condition at the electrode in question. The positive direction is denoted as the noble direction and the negative direction (active direction) is associated with reduction.

sign convention for electrode potential temperature coefficients—for both isothermal temperature coefficients and the thermal coefficients, the sign convention is that the temperature coefficient is positive when an increase in temperature produces an increase in the electrode potential.

significant digit—any digit that is necessary to define a value or quantity.

significant figures—number of digits used in a number to specify its accuracy.

silane (silicane)—a colorless gas (SiH_4) that is insoluble in water. Also any of a class of compounds of silicon and hydrogen having the general formula Si_nH_{2n+2}.

silica—common oxide of silicon; the compound silicon dioxide (SiO_2).

silica gel—rigid gel that is a porous silicon dioxide (SiO_2) made by coagulating a sol of sodium silicate and heating to drive off water. Used as a drying agent because it readily absorbs moisture from the air.

silicane—*see* **silane**.

silica reduction—removal of excess silica from raw water for direct use as cooling tower makeup, or as the first stage treatment followed by ion-exchange for boiler makeup or process use. Accomplished by sorption of the silica on magnesium hydroxide precipitate.

silicate—any of a group of substances containing negative ions composed of silicon and oxygen. Silicates consist of the orthosilicates (SiO_4^{4-}), pyrosilicates ($Si_2O_7^{6-}$), metasilicates (SiO_3^{2-}), and disilicates ($Si_2O_5^{2-}$).

silicate paint—paint employing a soluble silicate as a binder. Used primarily in inorganic, zinc-rich coatings.

silicide—compound of silicon with a more electropositive element. React with mineral acids to form silanes.

silicon carbide—compound of the composition SiC. Extremely hard, it is widely used as an abrasive. Also known as Carborundum, which is a trademark for a commercial abrasive of this basis.

silicon dioxide (silica)—silicon (IV) oxide (SiO_2).

silicone—family of polymeric materials containing chains of silicon atoms alternating with oxygen atoms and with the silicon atoms linked to organic groups. Siloxanes are the monomers of silicones.

silicone resin—resinous form of a silicone. They are used as film formers (binders) in water-repellent and high-temperature paints.

siloxanes—compounds containing silicon atoms bound to oxygen atoms with organic groups linked to the silicon atoms, e.g., $R_3SiOSiR_3$. Silicones are polymers of siloxanes.

silt—sedimentary material of fine mineral particles intermediate in size between sand and clay and that exhibit little or no swelling characteristic.

silver migration—form of electrolysis in which silver ions dissolve from one conductor in an electronic circuit and, under the impetus of a direct current and presence of moisture, migrate across an insulator to a second conductor.

simplest formula—*see* **empirical formula**.

SIMS—*abbreviation for* **secondary ion mass spectroscopy**.

simulated service corrosion test—term used with reference to engine coolants. It is an evaluation of the effects of a circulating engine coolant on metal test specimens and automotive cooling system components under controlled, essentially isothermal, laboratory conditions.

sintering—process of heating a compacted powdered material at a temperature below its melting point in order to weld the particles together into a single rigid shape.

slag—term used in metallurgical processing for the impurities separated from molten metal during its refining and that floats on the surface where it can be separated from the molten metal.

slaked lime—calcium hydroxide ($Ca(OH)_2$) or quicklime that has been slaked by addition of water. *Also called* **hydrated lime**. *See also* **calcium hydroxide**.

slaking—deterioration of rock on exposure to air or water. Also to combine chemically with water or moist air.

Sligh oxidation test—test for determining the oxidation resistance of oils.

slime—deposits formed on surfaces from the abundant growth and development of algae, fungi, and bacteria.

sling psychrometer—psychrometer (hygrometer) containing independently matched dry- and wet-bulb thermometers suitably mounted for swinging through the atmosphere for simultaneously indicating the dry- and wet-bulb temperatures that enable the relative humidity to be calculated.

slow drying—refers to paint films requiring a drying time 24 hours or longer before recoating.

slow strain rate test (SSRT)—test where the tensile specimen is subjected to a steadily increasing stress at constant strain rate in a given environment. Procedure results in rupture of surface films to eliminate the initiation time required for a surface crack to form.

sludge—water-formed sedimentary deposit. Also used to describe insoluble material formed as a result either of deterioration reactions in an oil or of contamination of an oil, or both. Also used to describe a boiler deposit that is softer and less adherent than scale.

sludge volume index—inverse measure of sludge density.

slurry—mixture consisting of a suspension of solid material in liquid of such consistency as to be capable of being pumped like a liquid.

slushing—process by which a coating is liberally applied to a surface by swilling or flooding with the excess drained off. Used to impart coating protection for surfaces that are not readily accessible for painting by ordinary methods.

slushing compound—nondrying oil, grease, or wax applied to metals for temporary corrosion protection. The oily equivalent is called slushing oil.

slushing oil—*see* **slushing compound**.

small-amplitude cycle voltammetry (SACV)—technique for measuring corrosion and polarization resistance.

SME—abbreviation for the Society of Manufacturing Engineers.

smelt—(1) molten slag. (2) cooking chemicals tapped from the recovery boiler in the pulp and paper industry as molten material and dissolved in the smelt tank. *See also* **green liquor**.

smelting—process of separating a metal from its ore by heating to a high temperature in the presence of a reducing and a fluxing agent.

SMIE—*abbreviation for* **solid metal induced embrittlement**.

smog—fog that has become mixed and polluted with smoke. Also the irritating haze resulting from the effect of the sun on certain pollutants in the air.

smoke—fine suspension of solid particles in a gas. Also carbon or soot particles less than 0.1 micrometer in size that result from the incomplete combustion of carbonaceous materials. Also an aerosol of particles.

smut—residue left behind after steel or cast iron is acid pickled. The most significant component of ferrous alloy smut is ferric carbide with some iron oxide and finely divided carbon. Smut is insoluble in sulfuric acid but soluble in hydrochloric acid. A smut also develops during the acid etching of aluminum due mainly to the alloying elements present.

SNAME—abbreviation for the Society of Naval Architects and Marine Engineers.

SNG—*abbreviation for* **synthetic (or substitute) natural gas**.

SO$_x$—term used for the anthropogenic (man-caused) and natural emissions of sulfur oxides into the atmosphere as pollutants that contribute to atmospheric corrosivity. The main part of anthropogenic SO$_x$ pollution is caused by combustion of fossil fuels emitted as gaseous SO$_2$, which is shortly (½ to 2 days) oxidized on moist particles or in droplets of water to sulfuric acid (H$_2$SO$_4$).

soap—cleaning agent usually consisting of sodium or potassium salts of fatty acids. Also the metallic salt of a long-chain organic carboxylic acid. Also the product formed by the saponification or neutralization of fats, oils, waxes, rosins, or their acids with organic or inorganic bases.

soapstone—mineral from which talc is obtained.

soda—sodium oxide (Na$_2$O). Also loose term for describing any of a number of sodium compounds such as caustic soda (sodium hydroxide, NaOH), and washing soda (sodium carbonate, Na$_2$CO$_3$·10H$_2$O).

soda ash—sodium carbonate (Na$_2$CO$_3$).

soda-base grease—grease prepared from a lubricating oil and a sodium soap.

soda lime—mixed hydroxide of sodium and calcium made by slaking lime with caustic soda solution and recovering the product by evaporation. Also a mixture of CaO and Na$_2$O used as a desiccant.

sodium carboxymethylcellulose—*see* **CMC**.

sodium cycle softening (sodium zeolite softening)—water softening process utilizing resin materials that have the property of exchanging the hardness constituents, calcium and magnesium, for sodium.

sodium hexametaphosphate—chemical compound of composition NaPO$_3$.

sodium orthophosphate—*see* **trisodium phosphate**.

sodium sesquicarbonate—double salt ($Na_2CO_3 \cdot NaHCO_3 \cdot 2H_2O$), which is less alkaline than sodium carbonate (Na_2CO_3) alone and has application as a water-softening agent.

sodium tripolyphosphate (STPP)—condensed phosphate chain containing the P–O–P group and represented by the formula $Na_5P_3O_{10}$. Used in water treatment as a sequesterant to prevent iron, manganese, and calcium carbonate deposits.

softening—removal of hardness (dissolved calcium and magnesium salts) from water.

soft iron—form of iron that contains very little carbon, has high relative permeability, is easily magnetized and demagnetized, and has a small hysteresis loss.

soft radiation—ionizing radiation of low penetrating power.

soft rot—*see* **surface rot**.

soft water—water free of or containing very low concentrations of dissolved calcium and magnesium salts.

softwoods—botanical group of trees having needlelike or scalelike leaves; generally conifers, but also includes cypress, larch, and tamarack. Term has no reference to the hardness of the wood. *See also* **hardwoods**.

soil—(1) solid contaminating foreign matter embedded in or adhered on a surface. Soils can be classified as organic or inorganic. Organic soils are generally oily, waxy films that are usually removed by alkaline cleaners. Inorganic soils include rust, smut, scale, and inorganic particulates and are usually removed by acidic cleaners. (2) the layer of unconsolidated particles derived from weathered rock, organic matter, water, and air that forms the upper surface over much of the earth and that supports plant growth.

soil corrosion—corrosion of underground metallic structures. It is an electrochemical corrosion process that can be prevented by isolating the metallic structure from contacting soil or by utilizing cathodic protection. Main factor causing soil corrosion is differential aeration of the soil.

soil stress—stress on coatings due to movement of contacting soil causing the coating to pull and creating a potential for cracks, voids, or thin spots.

sol—*abbreviation for* **hydrosol**. Also general term for a dispersion of a solid in a liquid. *Also abbreviation for* **colloidal solution**.

solar degradation—deterioration produced by exposure to solar energy.

solar energy—electromagnetic energy radiated from the sun.

solar heating—form of domestic or industrial heating relying on the direct use of solar energy.

solder—any alloy used to join metal surfaces. Commonly, solders are composed of lead and tin, often with small amounts of other elements present. Brazing solders are usually alloys of copper and zinc, which melt at over 800 °C (1472 °F). Soft solder melts in the range 200 to 300 °C (392 to 572 °F) and is a tin-lead alloy. Hard solder contains substantial quantities of silver in the tin-lead alloy.

solder bloom corrosion—corrosion mode of internal solder joints of an automotive radiator. Describes a flowering appearance of corrosion by-products that may lead to blockage of radiator tube passages and possibly result in engine overheating.

solder embrittlement—reduction in mechanical properties of a metal as a result of local penetration by solder along grain boundaries.

solid—state of matter in which there is a three-dimensional regularity of structure, resulting from the proximity of the component atoms, ions, or molecules and the strength of the forces between them. A solid has definite shape and volume in contrast to a liquid or gas.

solid lubricants—special types of extreme-pressure agents that are used separately or as insoluble additives to lubricating oils or greases. They do not have to react with the opposing metal surfaces to form a protective and lubricating film. Solid lubricants include molybdenum disulfide, graphite, polyfluorohydrocarbon solids, borates, phosphates, polyamides, and various glasses.

solid metal induced embrittlement (SMIE)—embrittlement of a solid metal occurring below the melting temperature of the solid in certain liquid-metal embrittlement (LME) couples where the solid is intimately contacted by the embrittler, there is tensile stress, a crack nucleation at the solid/embrittler interface, and a presence of the embrittling species at the propagating crack tip.

solids—paint term denoting the nonvolatile matter in a coating composition.

solid solution—single, solid, homogeneous crystalline phase containing two or more chemical species.

solid-solution hardening—process for hardening metals and alloys by adding one element to another to produce a solid solution.

solids volume—percentage of total volume of a paint occupied by nonvolatiles.

solubility—quantity of solute that dissolves in a given quantity of solvent, under specified conditions, to form a saturated solution.

solubility product—product of the activities (or concentrations as a close approximation) of ions in a saturated solution.

solubilization—any process used to increase the solubility of deposit-forming materials, e.g., by pH-lowering, sequestration, etc.

soluble anode—anode used for the corrosion protection of susceptible structures and that is significantly soluble in the electrolyte. Most commonly the soluble anode is scrap iron and old rail or pipe.

soluble cutting oil—*see* **soluble oil**.

soluble oil—oil-rich concentrate that will mix with water to form an emulsion imparting such properties as lubrication, cooling, and corrosion protection during metalworking operations. *See also* **metalworking fluids** and **semisynthetic fluids**.

solute—substance that is dissolved in a solvent to form a solution.

solution—in a chemical system, it is a single phase existing over a range of compositions. Also a homogeneous mixture of several components that is a single phase having identical properties throughout. It is composed of a solvent, which is the dissolving medium, and one or more solutes, which are the components that dissolve.

solution annealing—*see* **quench annealing**.

solution corrosion—same as swelling corrosion, but where new, readily soluble compounds are formed on the surface of the material.

solution potential—electrode potential where the half-cell reaction involves only the metal electrode and its ion.

solvation—interaction of ions of a solute with the molecules of a solvent. Also adsorption of a microlayer or film of water on individual dispersed particles of a solution or dispersion.

solvent—liquid part that dissolves the solute. Also a liquid in which another substance may be dissolved.

solvent-borne coating—coating containing only organic solvents. *Compare with* **water-borne coating**.

solventless paints—paints based on liquid resins of low viscosity and used where solvents are undesirable. A 100% solids coating system.

solvent wash—cleaning with a solvent.

solvolysis—reaction between a compound and its solvent, e.g., hydrolysis.

soot—agglomeration of particles of carbon impregnated with tar formed in the incomplete combustion of carbonaceous material.

sorption—taking up of gas by adsorption, absorption, chemisorption, or any combination of these. Often used when the specific mechanism is not fully known.

sour—natural gas containing such amounts of compounds of sulfur (mainly hydrogen sulfide, H_2S) as to make it impractical to use without purification because of its toxicity or its corrosive effect on piping or equipment.

sour corrosion—corrosion caused by hydrogen sulfide dissolved in water.

sour crude—crude oil that is corrosive when heated and evolves significant amounts of hydrogen sulfide (H_2S) on distillation. Usually, but not necessarily, the crude has a high sulfur content. Also called sour oil.

sour gas—gaseous environment containing hydrogen sulfide (H_2S) and usually also carbon dioxide (CO_2) found in hydrocarbon reservoirs.

souring—*see* **fermentation**.

sour water—wastewaters containing malodorous materials, usually sulfur compounds.

spalling—separation of a surface layer caused by thermal or mechanical stresses.

spark test—method of detecting holidays on metallic substrates by means of a spark test tool.

spar varnish—varnish having good water resistance for use on exterior surfaces.

SPE—abbreviation for the Society of Plastics Engineers.

specification—document setting forth a precise statement of a set of requirements to be satisfied by a material, product, system, or service.

specific bonding—chemical bond formed when an adhesive reacts chemically with the adherend.

specific conductivity—term allowing for comparison of the electrical conductivity of different electrolyte solutions. Its dimensions are reciprocal ohm centimeters.

specific gravity—ratio of the density of the substance to the density of a reference substance, usually water, at specified temperature and pressure.

specific gravity, apparent—ratio of the weight in air of a unit volume of material at a stated temperature to the weight in air of equal density of an equal volume of gas-free distilled water at a stated temperature.

specific heat—amount of heat required to raise the temperature of 1 gram of substance by 1 K.

specimen—individual unit on which a test or examination is made.

spectral energy distribution (SED)—general term for the characterization of the amount of radiation present at each wavelength and usually expressed by power in watts, irradiance in watts per square centimeter, or energy in joules. *See also* **irradiance** and **spectral irradiance**.

spectral irradiance—distribution of irradiance as a function of wavelength and expressed in watts per square meter per wavelength band. Spectral irradiance is the proper method for comparing sources with different energy distributions. *See also* **irradiance** and **spectral energy distribution**.

spectrophotometer—photometric device for the measurement of spectral transmittance, spectral reflectance, or relative spectral emittance.

spectroscopy—methods for producing and analyzing spectra using spectroscopes, spectrometers, spectrographs, and spectrophotometers. Infrared, visible, and ultraviolet spectroscopies are used for the qualitative and quantitative analysis of organic compounds and alloys. Raman spectroscopy can be used to characterize and identify inorganic, organic, and biological molecules and determine their structures. X-ray diffraction can be used to identify elements by their crystal structure. Atomic absorption spectroscopy can be used for the quantitative analysis of organic and inorganic materials and for metals. Emission spectroscopy can be used for the simultaneous analysis of all metallic elements in a sample. Mass spectroscopy is used to analyze samples according to their mass and therefore the technique is ideal for isotope and molecular weight determinations.

spectrum—spatial arrangement of components of radiant energy in order of their wavelengths, wave number, or frequency.

spent caustic—spent aqueous alkaline solutions that had been used to wash refinery gases and light refinery products. Usually classified as being sulfidic or phenolic and that may contain varying amounts of sulfides, sulfates, phenolates, naphthenates, sulfonates, mercaptides, and other organic and inorganic compounds.

sperm oil—pale-yellow liquid organic substance derived from the head cavities and blubber of the sperm whale. It is a liquid wax rather than fatty oil and is an excellent boundary lubricant highly resistant to gumming and viscosity decrease when subjected to high temperatures and pressures.

spheroidizing—treatment of carbon steel by heating at approximately 649 °C (1200 °F) for 24 hours or more to coalesce the iron carbide (Fe_3C) into its more stable globular form.

spindle oil—any low-viscosity mineral oil.

spinel—group of oxide minerals that crystallize to have a general composition that is a combination of bivalent and trivalent oxides of magnesium, zinc, iron, manganese, aluminum, and chromium.

spirit—paint term describing commercial ethyl alcohol. Differs from mineral spirits.

spirit varnish—varnish that hardens by solvent evaporation with no oxidation or polymerization occurring.

spit—expectorated saliva. *See also* **saliva**.

splash zone—marine exposure term defined as the zone extending from just below the lowest astronomical tide up to a height at which salt spray effects are still considered significant. It thus includes the whole tidal zone plus an atmospheric zone.

sponge iron—coherent, porous mass of substantially pure iron produced by solid-state reduction of iron oxide.

spontaneous passivation—metal condition that, at its primary passive potential, has its cathodic reduction rate equal to or greater than its anodic dissolution rate. *See also* **primary passivation potential**.

spore—asexual, usually single-celled, reproductive organ characteristic of nonflowering plants such as fungi, mosses, or ferns.

spot repair—preventive maintenance repainting of small areas.

spotting out—delayed appearance of spots and blemishes on plated or finished surfaces that is most prevalent on porous basis metals or substrates.

spray coating—aerosol spray product for surface application that leaves a residual finish for protective or decorative purposes.

spray pond—use of one or more spray nozzles in conjunction with a cooling pond to admit recirculating cooling water thereby increasing the rate of evaporation for faster cooling. *See also* **cooling pond**.

spreading rate—recommended area to be covered by a unit volume of paint and frequently expressed as square feet per gallon of paint.

spring—point at which groundwater under pressure leaves the ground to flow on the surface.

sputtering—process by which some of the atoms of an electrode, usually a cathode, are ejected as a result of bombardment by heavy positive ions. Used to clean a surface or to deposit a uniform film of a metal on another surface in an evacuated chamber.

squeeze treatment—method of continuously feeding a corrosion inhibitor into an oil well by pumping a quantity of the inhibitor into the well followed by sufficient solvent to force the inhibitor into the formation.

SRB—*abbreviation for* **sulfate-reducing bacteria**.

SSPC—abbreviation for the Steel Structures Painting Council.

SSRT—*abbreviation for* **slow strain rate test**.

stability index—*same as* **Ryzner index**.

stabilization—production of a water that is exactly saturated with calcium carbonate ($CaCO_3$).

stabilized steel—stainless steel that has been alloyed with a carbide-forming element such as titanium, niobium, or tantalum to make it less susceptible to carbide precipitation.

stabilizer—ingredient added to paint or plastic to retard possible degradation. Also any substance used to inhibit a chemical reaction.

stable—term used to describe substances that do not tend to undergo spontaneous changes.

stacking faults—those defects present in the normal sequence of stacking of atomic layers in metal crystal structures. Possible faults include: a slip plane where the lattice plane separating two regions of a crystal have slipped relative to each other; dislocations, which are defects existing in nearly all real crystals, caused by such as an additional layer of atoms or a staking fault on one side of a defect; point defects, which are vacancies caused by the absence of one or more atoms in the crystal, by impurity atoms, and by interstitial atoms; and grain-boundary defects.

stagnant—liquid, principally water, that is not moving or flowing and that can lead to concentration cell corrosion of contacting metal due to the difference in its low oxygen concentration compared with higher oxygen in nearby dynamic fluid resulting in an oxygen crevice corrosion cell.

stain—discoloration by foreign matter. Also refers to a transparent or semitransparent solution or suspension of coloring matter in a vehicle designed to color a surface by penetration without hiding it.

stainless steels

stainless steels—steel alloys containing at least 11 to 12% of chromium, a low percentage of carbon, and often some other elements, notably nickel and molybdenum. The steels are classified into four groups; martensitic, ferritic, austenitic, and age-hardened or precipitation-hardened. Stainless steels do not readily rust or stain and therefore have a wide variety of uses in industrial, chemical, and domestic environments.

stainless steel classification—martensitic (group I), ferritic (group II), austenitic (group III), and age-hardened or precipitation-hardened (group IV) are the four classes of stainless steels based on their chemical composition and structure.

stainless steel sensitization—impoverishment of protective chromium in the region adjacent to the grain boundaries of austenitic stainless steels resulting from the precipitation of chromium carbide on heating of the alloys between about 500 and 800 °C (932 and 1472 °F). The alloy impoverishment may reach chromium values below the required 11 or 12% needed for corrosion resistance and lead to preferential corrosion along grain boundaries, ultimately resulting in metal disintegration.

stamping—process used to cut lines of configurations, letters, figures, and decorations on smooth metal surfaces. Also the collective term for specialized stamping operations known as coining, embossing, blanking, and pressing.

standard—specification, practice, or test method formally adopted by an authorizing body. Also a reference used as a basis for comparison or calibration.

standard atmospheric conditions—29.92 inches mercury, 68 °F, and 30% relative humidity.

standard cell—any voltaic cell used as a standard of emf.

standard conditions—general term for describing the usual, imposed, or a standardized environment for testing. The ISO standard conditions are a temperature of 23 ± 2 °C and relative humidity of $50 \pm 5\%$.

standard deviation—*see under* **statistical terms**.

standard electrode—electrode used in measuring electrode potential.

standard electrode potential—reversible potential of an electrode process when all reactants and products are at unit activity on a scale in which the potential for the standard hydrogen half-cell is zero. It is important in theoretical considerations, but of limited practical value.

standardization—correlation of an instrument response to a standard of known accuracy.

standard solution—solution of known concentration for use in volumetric quantitative analysis.

standard state—state of a system used as a reference value in thermodynamic measurements. The standard states involve a reference value of pressure (usually 1 atmosphere or 101.325 kPa) or concentration (usually one molar). Thermodynamic functions are designated as standard when they refer to changes in which reactants and products are all in their standard and normal physical state.

standard temperature and pressure (STP)—standard conditions used when comparing the properties of gases and as a basis for calculations involving quantities that vary with temperature and pressure. These conditions are 273.15 K (or 0 °C) and 101,325 Pa (or 760.0 mm Hg).

starch—polysaccharide consisting of various proportions of two glucose polymers.

static electricity—effects produced by electric charges at rest, including the forces between charged bodies and the field they produce.

static friction—force just sufficient to initiate relative motion between two bodies under load.

statistic—summary value calculated from the observed values in a sample.

statistical terms—terms used in presenting precision and accuracy information. They include the following: (a) **error:** any deviation of an observed value from the true value, (b) **random error:** chance variation encountered in experimental work despite the closest possible control of variables, (c) **bias:** constant or systematic error as opposed to random error, (d) **precision:** degree of agreement of repeated measurements of the same property, (e) **accuracy:** agreement between an experimentally determined value and the accepted reference value, (f) **variance:** measure of the dispersion of a series of results around their average, (g) **standard deviation:** measure of the dispersion of a series of results around their average, expressed as the square root of the quantity obtained by summing the squares of the deviation from the average of the results and dividing by the number of observations minus one, (h) **coefficient of variation:** measure of relative precision calculated as the standard deviation of a series of values divided by their average, (i) **range:** absolute value of the algebraic difference between the highest and lowest values in a set of data, (j) **duplicates:** paired determinations performed by one analyst at essentially the same time, (k) **95% confidence limit:** interval or range of values around an observed value which will, in 95% of the cases, include the expected value, (l) **95% confidence level:** probability of a precision statement that there are 95 in 100 chances of being correct, and 5 in 100 chances of being wrong in predicting that the expected value will fall within the specified limits or range, (m) **repeatability:** precision of a method expressed as the agreement attainable between independent determinations performed by a single analyst using the same apparatus and techniques, (n) **reproducibility:** precision of a method expressed as the agreement attainable between determinations performed in different laboratories.

statistics—branch of mathematics concerned with the inferences that can be drawn from numerical data on the basis of probability.

stator—stationary electromagnetic structure of an electric motor or generator. *See also* **rotor**.

steam—(1) vapor phase of water. (2) mist of cooling water vapor.

steam boiler—closed tank in which water is converted into steam under pressure.

steam clean—cleaning process using pressurized steam or pressurized steam with added detergent.

steam condensate—water condensed from steam.

steam flooding—secondary oil recovery of low-gravity oil from relatively shallow formations by introducing a steam-hot water mixture from a once-through steam generator into one of a group of centrally located wells to radiate the steam and hot water outward. The oil thus treated is then recovered by pumping.

steam genny—steam generator used in the field to clean metals, coatings, or concrete surfaces. Detergents may be added to the water reservoir to provide for emulsification of any surface oils or organic contaminants.

steam-injected cleaning—chemical cleaning method for process equipment and piping where a concentrated mixture of cleaning chemicals are injected into a stream of fast-moving steam at one end of the system and condensed at the other. The steam atomizes the chemicals, increasing their effectiveness by ensuring good contact with metal surfaces.

steam metal—red brass type alloy whose nominal composition is 87% copper, 3% zinc, 7% tin, and 3% lead.

steam point—temperature at which the vapor pressure of water is equal to the standard atmospheric pressure, i.e., at 1 atmosphere pressure and 100 °C (212 °F).

steam purity—refers to the amount of solid, liquid, or vaporous contamination of steam from a boiler water system.

steam quality—a measure of the amount of moisture in steam from a boiler water system.

steel—any of a number of alloys consisting predominantly of iron with varying proportions of carbon (up to 1.7%) and, in some cases, small quantities of other elements that then are called alloy steels.

steel grades—used when classifying steel by carbon content. Low-carbon steel: 0.15% carbon maximum; medium low-carbon steel: 0.15% to 0.23% carbon; medium high-carbon steel: 0.23% to 0.44% carbon; high-carbon steel: greater than 0.44% carbon.

steel wool—matted mass of long, fine steel fibers available in a variety of grades of coarseness that are used in cleaning and polishing surfaces, removing blemishes, and dulling coated surfaces.

STEL—*abbreviation for* **Short-Term Exposure Limit**.

stepwise cracking—*same as* **blister cracking** and **hydrogen-induced cracking**.

steric effect—when the rate or path of a chemical reaction depends on the size or arrangement of groups in a molecule.

steric hindrance—when a chemical reaction is slowed or prevented because large groups on a reactant molecule hinder the approach of another reactant molecule.

sterile—free from any viable microorganisms, either active or dormant.

sterling silver—silver containing no more than 7.5% copper. The copper is used for hardening.

Stiff-Davis index—index for predicting the stability of brackish waters used in oil field waterflooding.

STLE—abbreviation for the Society of Tribologists and Lubrication Engineers.

STM—abbreviation for scanning tunnelling microscope.

Stoddard solvent—mineral spirits with a minimum flashpoint of 37.8 °C (100 °F), low odor level, otherwise conforms to Stoddard solvent specifications (ASTM D 484).

stoichiometric—pertaining to weight relations in chemical reactions. Also a substance having the precise weight relationship of its elements as those of a chemical compound.

stoichiometric compound—compound in which atoms are combined in exact whole-number ratios. *Opposite of* **nonstoichiometric compound**.

stoke—standard unit of kinematic viscosity in the c.g.s. system; expressed in cm^2/sec.

Stokes' law—law that predicts the frictional force on a spherical ball moving through a viscous medium. Used to calculate rate of fall of particles through a fluid based on densities, viscosity, and particle size.

stone—crushed or naturally angular particles of rock.

stoneware—heavy, nonporous pottery having good corrosion resistance and used for pipes, valves, pumps, sinks, crocks, and other vessels.

stop-leak additive—compound containing particulates that is added to automotive and truck cooling system coolants for the purpose of stopping or minimizing leaks.

storage life—*see* **storage stability** and **shelf life**.

storage stability—ability of a product to maintain its original characteristics over extended storage periods.

STP (stp)—*abbreviation for* **standard temperature and pressure**.

STPP—*abbreviation for* **sodium tripolyphosphate**.

straight mineral oil—petroleum oil containing no additives.

strain—dimensionless quantity, usually presented in %, inches per inch, or millimeters per millimeter for the change in the size or shape of a body due to applied force (stress). It is a measure of the extent to which a body is deformed when subjected to a stress.

strain hardening—*see* **work hardening**.

stratosphere—that region of the atmosphere next to the troposphere and located approximately from 12 to 50 kilometers above the surface of the earth. Its lower reaches contain the ozone layer.

stray current—electric currents following paths other than the intended circuit. Also direct current flowing in the earth and capable of causing corrosion damage.

stray-current corrosion—corrosion that is caused by stray currents from some external source, e.g., by any extraneous current in the earth.

streamline flow—fluid flow in which no turbulence occurs and the particles of the fluid follow continuous paths, either at constant velocity or at a velocity that alters in a predictable and regular way. *See also* **laminar flow**.

Streicher test for stainless steels—polished alloy specimen is etched in 10% oxalic acid for 1.5 minutes under an applied current density of 1 A/cm^2 and the surface then examined at 250 to 500×. Used to indicate sensitized stainless steel having intergranular corrosion susceptibility. Details of the test are given in ASTM A 262-55T.

stress—load or force per unit area of the cross section through which the load is acting. Also the resistance to deformation developed within a material subjected to an external force.

stress-accelerated corrosion—increased corrosion caused by applied stresses.

stress cell—corrosion cell developed in a metal or alloy where one section is more highly stressed than another. The higher stressed section is anodic to the lower stressed section.

stress-corrosion cracking—metal or alloy cracking process that requires the simultaneous action of a corrodent and sustained tensile stress.

stress relieving—heating a material to a suitable temperature and holding it long enough to reduce residual stresses, then cooling slowly enough to minimize the development of new residual stresses.

stress-sorption cracking—name given to a proposed mechanism for stress-corrosion cracking that concludes that it does not proceed by electrochemical dissolution of metal but by weakening of the cohesive bonds between surface metal atoms through adsorption of damaging components of the environment.

stress-strain diagram—diagram in which corresponding values of stress and strain are plotted against each other.

strike—thin film of metal to be followed by other coatings. Also to plate for a short time, usually at a high initial current density.

strip coating—*same as* **coil coating**.

striping—edge painting prior to priming.

strippable coatings—coatings of synthetic polymers used for temporary protection or to "cocoon" equipment to prevent interchange of moisture with the atmosphere. Can readily be removed using organic solvents or by physical peeling.

stripper—device or chemical used to remove a coating from a substrate by means of pyrolyzing, chemicals, solvency, mechanical abrasion, or a combination of these.

strong acid—acid that is completely or nearly completely dissociated in aqueous solution. *Opposite of* **weak acid**.

strong base—base that ionizes or dissociates nearly completely in aqueous solution. *Opposite of* **weak base**.

strontium chromate—$SrCrO_4$. A bright yellow corrosion-inhibiting pigment used in primer paints.

stucco—exterior textured finish applied from a composition of Portland cement, lime, and sand, which are mixed with water.

styrene butadiene rubber (SBR)—synthetic rubber that is a copolymer consisting of approximately 25% styrene and 75% butadiene.

styrene plastics—plastics based on polymers of styrene or copolymers of styrene with other monomers.

subacute toxicity—property of a substance to cause adverse effects in an organism upon repeated or continuous exposure within less than the lifetime of that organism.

sublimate—solid formed by sublimation.

sublimation—conversion of a solid to a gaseous state bypassing the liquid phase.

substance—matter. Also material of specified constitution. *See also* **material**.

substitution reaction—chemical reaction where one atom or molecule is replaced by another atom or molecule. Also called displacement reaction.

substrate—basic surface. Material or object upon which something resides.

subsurface corrosion—formation of isolated products of corrosion beneath the material surface, usually due to selective reaction of certain alloy constituents. *Also called* **internal oxidation**.

sulfanes—compounds of hydrogen and sulfur containing chains of sulfur atoms. They have the general formula H_2S_n. The simplest member is H_2S, with others having progressively greater quantities of S.

sulfate attack—deleterious reactions between soluble sulfates in soil or groundwater and concrete or grouting. Characterized by scaling or powdering of the concrete or grouting due to expansive forces produced by the reaction product.

sulfate nests—localized concentration of ferrous sulfate near the metal surface of iron atmospherically rusting in sulfur oxide polluted environments. Its presence accelerates corrosion due to water absorption onto the surface, by providing the electrolyte to carry the external corrosion current, and by promoting a weakened, porous, nonprotective oxide surface.

sulfated oil—*same as* **sulfonated oil**.

sulfate-reducing bacteria (SRB)—anaerobic bacteria that reduce sulfate to sulfide. The usual by-product of this metabolic process is hydrogen sulfide (H_2S). The bacteria are widespread in soil, seawater, fresh water, and muddy sediments. They may cause anaerobic corrosion of iron and steel; when they are present, there is an abundance of sulfate, and the surface of the substrate is at between pH 5.5 and 8.5. The produced sulfide tends to retard cathodic reactions, particularly hydrogen evolution, and accelerates anodic dissolution thus increasing corrosion. The corrosion product is iron sulfide, which precipitates when ferrous and sulfide ions are in contact. *See also* **biological corrosion**.

sulfidation—oxidation by sulfur. Also reaction of a metal or alloy with sulfur-containing species to produce a sulfur compound forming on or beneath the surface.

sulfide attack—attack of copper alloys by cooling waters, usually brackish water or seawater, that are polluted with sulfides, polysulfides, or elemental sulfur. May increase general corrosion rates, and induce or accelerate dealloying, pitting, erosion-corrosion, intergranular attack, and galvanic corrosion.

sulfide stress cracking—hydrogen-stress cracking of a metal in an environment containing hydrogen sulfide (H_2S).

sulfidic corrosion—metal corrosion whose products are metal sulfides and most often encountered in petroleum refining and petrochemical operations. The corrodent is any of a variety of sulfur compounds made active at process temperatures between 260 and 540 °C (500 and 1000 °F). The sulfur compounds originate with crude oils and include polysulfides, hydrogen sulfide, mercaptans, aliphatic sulfides, disulfides, and thiophenes.

sulfinate (dithionite, hyposulfite)—salt that contains the negative ion $S_2O_4^{2-}$.

sulfite—salt or ester derived from sulfurous acid and containing the ion SO_3^{2-}. Used as a reducing agent and oxygen scavenger.

sulfite pulp—wood pulp produced by cooking with a sulfite liquor made by dissolving sulfur dioxide in an aqueous medium.

sulfonate—salt or ester that can ionize to form the sulfonate ion $SO_2 \cdot O^-$. *See also* **sulfonic acids**.

sulfonated oil—water dispersible or soluble surface active material obtained by treating an unsaturated or hydroxylated fatty oil, acid, or ester with an agent capable of sulfating or sulfonating it either wholly or partially. Also called sulfated oil.

sulfonation—chemical reaction in which a $-SO_3H$ group is substituted on a benzene ring to form a sulfonic acid. The reaction is commonly carried out by refluxing with concentrated sulfuric acid for a long time period.

sulfonic acids—organic compounds containing the $-SO_2OH$ group. They are strong acids, ionizing completely in solution to form the sulfonate ion.

sulfurized oil—oil to which sulfur or sulfur compounds have been added.

supercooling—metastable equilibrium condition of cooling a liquid below its freezing point without the separation of the solid phase.

supercritical steam plant—steam plant operating above the critical temperature, 375 °C (705 °F), and critical pressure, 22,000 kPa (3200 psi), of water.

Superfund—common designation for the U.S. Comprehensive Environmental Response, Compensation, and Liability Act.

superheated steam—steam that has additional heat added to it thus increasing its temperature and energy.

superheater—heat exchanger located in a furnace to increase the temperature of steam leaving the boiler drum.

superheating—heating of a liquid to above its normal boiling point by increasing the pressure.

supernate—liquid above the surface of settled sediment or precipitate.

superphosphate—commercial phosphate mixture that is more soluble than calcium phosphate ($Ca_3(PO_4)_2$) and that is used as a fertilizer; manufactured by treating calcium phosphate with sulfuric acid to yield $Ca(H_2PO_4)_2 \cdot 2CaSO_4$.

super refractories—high-melting-point metal oxides, carbides, and silicides that are generally very resistant to attack by molten metals, slags, and hot gases and extensively used for lining furnaces. Examples include magnesia, mullite, alumina, silicon carbide, and stabilized zirconia. *See also* **refractories**.

supersaturated solution—solution containing more than the maximum equilibrium (saturated) amount of solute at a given temperature. Supersaturated solutions are metastable.

supersaturated vapor—unstable condition where a vapor remains dry although its heat content is less than that of dry and saturated vapor at the same pressure. Found in steam emerging from the nozzle of a steam turbine. Phenomenon is similar to supercooling.

superstructure (topside)—area above the main deck of a ship where offices, staterooms, and equipment are located and that also includes the masts and antenna.

supplemental coolant additive (SCA)—chemical additive package that is added to heavy-duty diesel engine cooling systems to supplement corrosion protection of the engine coolant concentrate being used and to replace coolant additives lost because of dilution or depletion in service.

surface-active agent—material that when added to a liquid medium modifies the properties of the medium at a surface or interface. Also a general term describing the action of soluble detergents, dispersing agents, emulsifying agents, foaming agents, penetrating agents, and wetting agents.

surface alloying—*see* **diffusion coating**.

surface area—total area of the surface of any three-dimensional object, or of all particles in a mass of particles where it is usually expressed in terms of square meters per gram.

surface film cell—corrosion cell developed when one area of a metal or alloy surface develops a chemical or electrochemical reaction film while an adjacent area does not. The filmed area becomes cathodic to the clean area.

surface modification—alteration of surface composition or structure by the use of energy or particle beams (ion implantation or laser surface processing). Surface-modified layers are distinguished from conversion or other coating layers by their greater similarity to metallurgical alloying versus chemically reacted, adhered, or physically bonded layers.

surface rot (soft rot)—deterioration of the wood used in cooling towers and caused by cellulose-destroying fungi of the *Ascomycetes* and *Fungi imperfecti* groups leaving only the lignin binder and, consequently, little structural strength.

surface tension—contractile surface force of a liquid by which it tends to assume a spherical form and thus to present the least possible surface.

surface topography—geometrical detail of a solid surface, relating particularly to microscopic variations in height.

surface water—open body of water that is comprised of the lakes, rivers, or seas.

surfactant—contraction of the term surface-active agent.

suspension—two-phase system of finely divided solid particles dispersed in a solid, liquid, or gas.

SW 111—trademark for an anticorrosive pigment chemically described as calcium strontium phosphosilicate.

swab test—low-voltage electrical test for evaluating the continuity of porcelain enamel.

swarf—workpiece cuttings, dust, oil, grain particles, etc., created by any abrading action.

sweat—salty fluid secreted by sweat glands onto the surface of the skin.

sweating—condensing moisture on a surface.

sweet—refers to natural gas containing such small amounts of sulfur compounds that it can be used without purification with no toxic effects and no deleterious effect on piping and equipment.

sweet corrosion—corrosion resulting from the presence of water containing dissolved carbon dioxide, formic acid, or other short-chain acids in produced gas and/or oil fields.

sweet crude—crude oil that is not corrosive when heated and does not produce significant amounts of hydrogen sulfide (H_2S) on distillation. Also called sweet oil.

sweet oil/gas wells—production wells containing dissolved carbon dioxide but no or very little sulfur compounds. *See also* **sweet**.

swelling corrosion—used to describe the chemical reaction of aqueous solutions (natural and industrial waters, gases dissolved in moisture, etc.) to corrode building materials (nonmetallics and organic) by causing the formation of voluminous new compounds produced inside the material.

syndet—contraction for the term synthetic detergent.

syneresis—spontaneous separation of liquid from a gel.

synergism—combined action of several chemicals that produces an effect greater than the additive effects of each.

synfuels—liquid or gaseous fuels produced from coal, lignite, or other solid-carbon sources.

syntactic foam—lightweight system obtained by incorporating prefoamed or low-density fillers.

synthesis—formation of chemical compounds from simpler compounds.

synthetic—relating to or involving a man-made product.

synthetic fluids (chemical fluids)—metalworking fluids that are aqueous solutions of inorganic and organic rust inhibitors, extreme-pressure agents, and other soluble functional additives.

synthetic lubricant—lubricant produced by synthesis rather than by extraction or refinement of natural organic products.

synthetic (substitute) natural gas (SNR)—mixture of gaseous hydrocarbons produced from coal, petroleum, etc., and is suitable for use as a fuel.

synthetic oils—lubricants generally based on either synthetic hydrocarbons, organic esters, polyglycols, phosphate esters, silicon-containing compounds, halogen-containing compounds, halogenated polyaryls, fluorocarbons, or perfluoropolyglycols.

synthetic resin—*see* **resin, synthetic**.

systematic errors—correctable errors in measurement resulting from poor techniques and procedures, uncalibrated equipment, and human error.

T

tack—degree of stickiness.

Tafel slope—slope of the straight-line portion of a polarization curve, usually occurring at more than 50 millivolts from the open-circuit potential when presented in a semilogarithmic plot in terms of volts per logarithmic cycle of current density. The plot is known as the Tafel line, and the overall diagram is termed a Tafel diagram.

tailings—residue, coarse or fine, removed from a separation process.

talc—phyllosilicate mineral having the general formula $3MgO \cdot 4SiO_2 \cdot H_2O$. Noted for its softness, low thermal and electrical conductivity, and fire resistance.

tall oil—generic name for a number of products obtained from the manufacture of wood pulp by the alkali (sulfate) process, better known as the kraft process. Includes fatty acids, pitch, and rosin among others.

tall oil fatty acids—products generally containing 90% or more fatty acids obtained by fractionation of crude tall oil. *See also* **tall oil**.

tallow—animal fat prepared from beef and mutton.

tannin (tannic acid)—one of a group of complex, nonuniform organic chemicals commonly found in leaves, unripe fruits, and the bark of trees, notably oak. Chemically, they may be divided into two groups: (a) derivatives of flavanols, and (b) hydrolyzable esters of a sugar with one or more trihydroxybenzenecarboxylic acids. The latter find use as rust converters.

tape—any continuous, narrow, flexible strip of material, usually of polyvinyl chloride or polyethylene basis, used as a corrosion-protective measure (barrier protection) for underground piping.

tape-pull test—cleanliness test for the effectiveness of a cleaning operation in which any remaining inorganic soil on the surface is made more easily visible. Transparent adhesive tape is applied to the dried surface then removed and placed on a white piece of paper for viewing. The contrast afforded by the white paper background allows for easy identification of any remaining soil.

tape test—adhesion test consisting of applying an adhesive tape to a dried coating, either scribed or unscribed, and rapidly removing the tape with a swift, jerking motion. The amount of coating removed is a measure of adhesion.

tapping—machining process for making internal threads using a tap which is a cylindrical or conical thread-cutting tool having threads of a desired form on the periphery. The internal thread is formed by the tap cutting with a combined rotary and axial motion.

tar—brown to black bituminous material (liquid or semisolid) in which the predominating constituents are bituminous and being a mixture of hydrocarbons and carbon. Obtained as condensate in the processing of coal, petroleum, oil shale, wood, or other organic materials.

tarnish—surface discoloration of a metal caused by a thin film of corrosion product.

tar sands (oil sands and bituminous sands)—sands (84 to 88%) having an organic component that is bitumen, a hydrocarbon mixture (8 to 12%), and water (4%). *See also* **bitumen**.

TCP—*abbreviation for* **tricresyl phosphate**.

TDS—*abbreviation for* **total dissolved solids**.

Teflon—trademark for a form of polytetrafluoroethylene.

telechelic polymers—polymers designed to contain terminal functional groups.

telomer—polymer composed of molecules having terminal groups incapable of reacting with additional monomers, under the conditions of synthesis, to form larger polymer molecules of the same chemical type.

TEM—*abbreviation for* **transmission electron microscopy**.

temper (draw)—to reheat hardened steel, normalized steel, or hardened cast iron to some temperature below the eutectoid temperature usually up to 100 to 200 °C (212 to 392 °F) for the purpose of decreasing hardness and increasing toughness. Tempering allows carbon to diffuse, causing the precipitation of small carbides.

temperature—thermal state of matter as measured on a defined scale. It is the degree of hotness or coldness of a body or an environment referenced to a specific scale. Usually expressed either as degrees Fahrenheit (F), Celsius (C), or kelvin (K). *See also* **dry-bulb temperature** and **wet-bulb temperature**.

temperature cell—corrosion cell developed in a metal or alloy where one section is maintained at a higher temperature than an adjacent section. The higher temperature section becomes anodic to the lower.

temperature indicating (TI) paints—paints applied to process equipment operating at elevated temperatures that give early warning, by means of their color change, of overheating.

temper color—thin, adherent oxide film that forms on steel tempered at a low temperature.

temporary coating (temporary protective)—coating designed to protect or decorate a substrate for a limited time, and one that can be readily removed by solvent, chemical, or mechanical means. *See also* **rust-preventive compounds**.

temporary hardness—main cause of natural water hardness. Water that contains dissolved calcium hydrogen carbonate ($Ca(HCO_3)_2$), which is formed in limestone or chalk regions by the action of dissolved carbon dioxide on calcium carbonate. It can be removed from the water by boiling, which results in the formation of insoluble calcium carbonate ($CaCO_3$), hence the name temporary hardness.

tenside—term for surface-active agent widely used in Europe.

tensile strain—relative length deformation by a specimen under tensile force.

tensile strength—maximum resistance of a material to deformation in a tensile test carried to rupture. Distinct from strength shown in compression, torsion, or shear.

tensile stress—stress that causes two parts of an elastic body, on either side of a typical stress plane to pull apart.

teratogen—toxic chemicals capable of causing birth defects.

ternary compound—compound containing three different elements.

terne—alloy of lead containing 3 to 15% tin and used as a hot dip coating for steel sheet or plate giving the steel better corrosion resistance and enhancing its ability to be formed, soldered, or painted.

terneplate (terne coating)—hot dipped coating of an alloy of lead and tin (*see also* **terne**) on a steel substrate.

terpenes—class of unsaturated organic compounds having the empirical formula $C_{10}H_{16}$ and occurring in most essential oils and oleoresinous plants.

terpolymer—polymer formed from three monomer species.

terrazzo—mosaic flooring.

test electrode—electrode whose potential is to be determined by comparison to a reference electrode.

test fence—fencelike structure strategically located in a part of the country for specific weather conditions (temperature, humidity, corrosivity, etc.) and facing a specific direction and angle. It contains a series of exposure racks on which test panels are affixed for exposure.

test method—definitive, standardized set of instructions for the identification, measurement, or evaluation of one or more qualities, characteristics, or properties of a material.

tetrasodium pyrophosphate (TSPP)—chemical compound of composition $Na_4P_2O_7$.

TFE fluorocarbon—*abbreviation for* tetrafluoroethylene or its polymer **polytetrafluoroethylene** $(C_2F_4)_n$.

the 20 ppb rule—seawater containing oxygen levels less than 20 ppb corrodes carbon steel at a rate of about only one mpy as long as the pH remains about 8 and velocities are less than 10 feet per second.

theoretical density—substance density calculated from the number of atoms per unit cell and measurement of the lattice parameters.

theory—unified set of hypotheses that are consistent with one another and with experimentally observable phenomena.

thermal analysis—technique for chemical analysis and the investigation of products formed by heating a substance.

thermal conductivity—time rate of heat flow under steady conditions through unit area per unit temperature gradient in the direction perpendicular to the area.

thermal decomposition—chemical breakdown of a compound or substance by temperature into simpler substances or into its constituent elements.

thermal diffusion—diffusion taking place in a fluid as a result of a temperature gradient.

thermal discharge—introduction of water from a source at a temperature different from the ambient temperature of the receiving water.

thermal electromotive force—electromotive force generated in a circuit containing two dissimilar metals when one junction is at a temperature different from that of the other. Used to make and utilize thermocouples.

thermal spraying—group of processes wherein finely divided metallic or nonmetallic materials are deposited in a molten or semimolten condition to form a coating.

thermal stability—resistance to permanent changes in property caused solely by heat.

thermal sterilization—steam-injection technique for killing wood-destroying fungi present in cooling tower wood. Procedure also provides for steam distillation of a fungicide onto the wood surfaces to minimize reinfection of the wood by fungi.

thermocouple—device consisting of two dissimilar metal wires or semiconducting rods welded together at their ends such that a thermoelectric emf is generated in the device when the ends are maintained at different temperatures and with the magnitude of the emf related to the temperature difference.

thermodynamics—study of the laws that govern the conversion of energy from one form to another, the direction in which heat will flow, and the availability of energy to do work.

thermodynamics laws—set of three laws, and an assumed other, that govern the conservation of energy from one form to another, the direction in which heat will flow, and the availability of energy to do work. The first law states that the sum of all the energies in an isolated system is constant, or as expressed mathematically, that for nonadiabatic systems of constant mass, the change in internal energy of the system (total kinetic and potential energy of the atoms and molecules of the system) equals the transfer of heat to or from the system plus the work being done on or by the system. The second law describes that all systems tend to approach a state of equilibrium that leads to the statement that it is impossible to transfer heat continuously from a colder system to a warmer system as the sole result of any process. Mathematically, it is expressed that the change in internal energy of the system equals the temperature of the system times its change in entropy. The third law states that for changes involving only perfect crystalline solids at absolute zero, the change of the total entropy is zero. This law enables absolute values to be stated for entropies. Another law, usually called the zeroth law of thermodynamics, is used in thermodynamics because it is assumed by the others. This law states that if two bodies are each in thermal equilibrium with a third body, then all three are in thermal equilibrium with each other.

thermoelectricity—electric current generated by temperature difference.

thermogalvanic corrosion—galvanic corrosion resulting from temperature differences at two points of a common metal surface in the presence of an electrolyte.

thermography (infrared imaging)—nondestructive technique for inspecting refractories, insulation, and furnace tubes to check surface temperature patterns as an indicator of their physical state and any local corrosion problems.

thermometer—instrument used for measuring the temperature of a substance.

thermopile—device used to detect and measure the intensity of radiant energy, consisting of a number of thermocouples connected together in series to achieve greater sensitivity.

thermoplastic—plastic that repeatedly can be softened by heating and hardened by cooling.

thermoplastic elastomers—range of organic materials with properties of rubber and plastic. Processed like conventional plastics in extrusion and injection-molding equipment, but have the flexural properties, tear strength, and abrasion resistance of rubber.

thermoplastic vulcanizates—mechanically compounded mixtures of polypropylene and a vulcanized rubber, such as ethylene-propylene-diene monomer.

thermoset—plastic that when cured by heat or chemical means cannot be resoftened by heat without decomposition.

thermostat—device that controls the heating or cooling of a substance in order to maintain it at a constant temperature.

thickeners—any material used to thicken a liquid. Also agents used to thicken lubricating oil to make a lubricating grease. Most commonly employed are the calcium, sodium, lithium, and aluminum salts of soaps, i.e., stearates and their mixtures, polyureas, and bentonite or hectorite clay.

thin-film lubrication—condition of lubrication in which the film thickness of the lubricant is such that the friction between surfaces is determined by the properties of the surfaces as well as by the viscosity of the lubricant.

thin-layer chromatography—technique for the analysis of liquid mixtures using chromatography where the stationary phase is a thin layer of an absorbing solid on a plate and the mixture to be analyzed is placed as a spot near one edge and the plate is stood upright in a solvent.

thinner—volatile organic liquid used to reduce paint viscosity. Also the portion of a paint, varnish, lacquer, or printing ink that volatilizes during the drying process.

thiols—organic compounds containing the group –SH (the thiol group). Also called mercaptans.

thixotropy—property of a material that enables it to stiffen in a relatively short time on standing, but on agitation or manipulation to change to a very soft consistency or to a fluid of high viscosity with the process being reversible.

thorough blast cleaning—*see* **white blasting**.

threading—cutting of external threads on cylindrical or tapered surfaces.

threshold effect—see threshold treatment.

threshold inhibitors—deposit control chemicals that are sequestering agents, such as polyphosphates and polyacrylates, that prevent formation of iron, manganese, and calcium carbonate deposits when used at far less than the concentrations required for softening on a stoichiometric basis.

Threshold Limit Value Concentration (TLV-C)—Time-Weighted Average (TWA) of the airborne concentration of a substance to which workers can be exposed for a normal 8-hour day, 40-hour work week without ill effects.

threshold treatment—term used in water treatment describing the control of scale or deposits by application of substoichiometric dosage of a treatment chemical. Also called threshold effect.

throwing power—relationship between the current density at a point on a surface and its distance from the counterelectrode. Also the ability of a plating solution to produce a uniform metal deposit on an irregularly shaped cathode.

tie coat—intermediate coat used to bond different types of paint coats.

timber—wood (lumber) used as a building material; usually a dressed piece of wood, especially a beam in a structure.

timber rotting—biochemical process taking place when the water content of the timber is in the range of about 20 to 60%, at temperatures between about 3 to 40 °C (37 to 104 °F), in stagnant air and relatively dark conditions. The action of fungi decomposes cellulose and hemicellulose yielding dry rot, wet rot, brown rot, cellar rot, etc.

timber weathering (timber corrosion)—surface effect (deterioration) on timber by the action of light, moisture, and oxygen where it takes on a gray tint.

time-of-wetness—refers to the effect of atmospheric moisture, principally relative humidity, on the corrosion rate of metal exposed outdoors. The rate is also dependent on the types and concentrations of aggressive pollutants in the atmosphere. Because meteorological fluctuation in most areas is repeated every year, the time-of-wetness in that location is almost constant every year.

time-proportioned sample—sample collected at preselected time intervals.

Time-Weighted Average (TWA)—air concentration of a chemical substance, for a normal 8-hour workday and 40-hour work week, to which all workers may be repeatedly exposed, day after day, without adverse effect.

tinning—coating of a metal with a very thin layer of molten solder or brazing filler material.

tin pest—phenomenon occurring with metallic pure tin subjected to temperatures below 20 °C (68 °F), at which it is metastable and may form an enantiomorphic gray tin that is a brittle material ultimately disintegrating into a sandy powder. Small alloying additions prevent the transformation.

tinplate—tin-coated steel primarily used in can making. The tin can be applied by dipping, but is mostly electroplated.

tin whiskers—filaments of tin that grow on tinplate, probably because of strains set up in the tinplate during the plating process, and that can grow long enough to short out adjacent circuitry when such tinplate is used in electronic hardware.

titration—method of volumetric analysis in which a volume of one reagent is added to a known volume of another reagent slowly from a burette until an end point is reached. The volume added before the end point is reached is noted. If one of the reagents has a known concentration, that of the other can be calculated.

titrimetric analysis—*see* **volumetric analysis**.

TKPP—abbreviation for tetrapotassium pyrophosphate ($K_4P_2O_7$).

TLV-C—*abbreviation for* **Threshold Limit Value Concentration.**

TOC—*abbreviation for* **total organic carbon**.

tolerance—total amount by which a quantity is allowed to vary.

toluol—commercial grade of toluene (phenylmethane, C_7H_{10}) of indeterminate purity.

ton of refrigeration—equivalent to the refrigeration produced by melting 1 ton of ice at a temperature of 0 °C (32 °F) in 24 hours.

tooth—profile of a surface; surface roughness. Also mechanical anchorage for a coating.

top coating—finish coat.

torr—unit of pressure equal to 1 mm Hg ($1/760$th of an atmosphere).

TOSCA (TSCA)—*abbreviation for* **Toxic Substances Control Act**.

total carbon—sum of the free carbon and combined carbon in a ferrous alloy.

total dissolved solids (TDS)—total amount of dissolved matter in a water. It is determined gravimetrically by evaporation of the water after removal of any suspended solids.

total hardness—sum of concentrations of all divalent cations (mostly calcium and magnesium ions) expressed as calcium carbonate ($CaCO_3$). U.S. Geological Survey hardness criteria for the total hardness of potable water is: soft water, 0 to 60 mg/L; moderately hard water, 61 to 120 mg/L; hard water, 121 to 180 mg/L; and very hard water, >180 mg/L $CaCO_3$.

total matter—sum of the particulate and dissolved matter in a water.

total organic carbon (TOC)—total organic carbon materials dissolved in water. Does not include carbonates, bicarbonates, or carbon dioxide.

total solids (TS)—sum of dissolved and suspended solids in a water. Determined gravimetrically.

total suspended solids (TSS)—total amount of insoluble particles suspended in a cooling water.

touch-up painting—painting of small areas of a previously painted surface to repair mars, scratches, and deteriorated coating in order to restore the coating to an unbroken condition.

toughness—property of a material by virtue of which it can absorb work. *Opposite of* **brittleness**.

toxic—poisonous.

toxicity—propensity of a substance to produce adverse biochemical or physiological effects.

Toxic Substance Control Act of 1977 (TSCA)—establishes inventory of chemicals by type, provides for filing a premanufacturing notice of new chemicals with the EPA, and provides for investigating, examining, and possibly placing restrictions on materials.

trace—constituent or impurity in a material making up only a very small portion of the sample.

trademark—name, symbol, or other device identifying a product officially registered and legally restricted to the use of the owner or manufacturer.

trade name—name by which a commodity, service, process, or the like is known to the trade. Also the name under which a business firm operates.

trade sales paints—paints, such as house paints, maintenance paints, and structure paints, which are applied on-site at ambient conditions by brushing or roller coating and which paints are usually purchased over-the-counter in relatively small quantity.

tramp oil—oil leakage from the hydraulic or lubrication systems of machine tools.

transference—movement of ions through an electrolyte and associated with passage of the electric current. *Also called* ion transport and **migration**.

transference number—*see* **transport number**.

transformer—device to increase or decrease (step-up or step-down transformer, respectively) the input voltage to a value required by the circuit receiving power.

transgranular corrosion—form of localized subsurface attack where a narrow path is corroded, at random, across the grain structure of a metal, disregarding grain boundaries.

transistor—semiconductor device capable of amplification in addition to rectification.

transition metal—metal element of groups IB through VIIIB of the periodic table.

transition primer—coating compatible with primer and also with a finish coat. Used where the latter is incompatible with the primer.

translucent—permitting the passage of radiation but not without some scattering or diffusion. *Compare with* **transparent**.

transmission—process by which radiant energy is transmitted through a material or an object.

transmission electron microscopy (TEM)—microscopic examination by allowing an electron beam to pass through a thin section (one micrometer or less) of the sample and projecting an image of the sample on a fluorescent screen. The diffraction pattern of the radiation may also be viewed to make phase identification possible.

transmission microscope—microscope where the image forming rays pass through the specimen being observed.

transmittance—fraction of emitted light of a given wavelength that is not reflected or absorbed, but passes through a substance.

transparent—permitting the passage of radiation without significant deviation or absorption. *Compare with* **translucent**.

transpassive region—region of an anodic polarization curve noble to and above the passive potential range where there is a significant increase in current density (metal dissolution) as the potential becomes more positive (noble).

transpassive state—state of an anodically passivated metal that is characterized by a considerable increase in the corrosion current and in the absence of pitting when the potential is increased.

transpiration—respiration of plants.

transport number (transference number)—that part (as a fraction) of the current carried by a certain ionic species in an electrolyte solution. Its symbol is t.

trees—branched or irregular projections formed on a cathode during electrodeposition, especially at edges and other high current density areas. Also tall woody plants, distinguished from shrubs by having comparatively greater height, and, characteristically, having a single trunk rather than several stems.

triboelectricity—static electricity produced as a result of friction.

tribology—science and technology of interacting surfaces in relative motion. Deals with friction, wear, and lubrication.

tricresyl phosphate (TCP)—$(CH_3C_6H_4)_3PO_4$. Commonly used plasticizer.

triethanolamine—$N(CH_2CH_2OH)_3$. Hygroscopic, viscous liquid with an ammoniacal odor. Used in metalworking fluids, manufacture of surfactants and emulsions.

trimer—polymer composed of molecules containing three identical mers.

triple bond—covalent bond in which six electrons are shared between two atoms.

triple point—temperature and pressure at which the vapor, liquid, and solid phases of a substance are in equilibrium.

tripoli—friable, dustlike silica used as an abrasive. *See also* **rottenstone**.

trisodium phosphate (sodium orthophosphate)—colorless crystalline compound (Na_3PO_4). Used as an additive in high-pressure boiler feed water for the removal of calcium and magnesium as precipitated phosphates, as a water-softening agent, corrosion inhibitor, and as a component in detergents and cleaning agent.

trivial name—vernacular name, or name for a substance which name is not produced by any systematic procedure.

trona—mineral form of sodium sesquicarbonate consisting of a mixed hydrated sodium carbonate and sodium hydrogencarbonate ($Na_2CO_3 \cdot NaHCO_3 \cdot 2H_2O$).

troposphere—layer of the atmosphere nearest Earth and extending approximately to 12 kilometers above the Earth.

true density—ratio of the mass of a material to its true volume.

true solution—a mixture (liquid, solid, or gaseous) in which the components are uniformly distributed throughout the mixture.

TS—*abbreviation for* **total solids**.

TSCA—*abbreviation for* **Toxic Substances Control Act**.

TSP—*abbreviation for* **trisodium phosphate**.

TSPP—*abbreviation for* **tetrasodium pyrophosphate**.

TSS—*abbreviation for* **total suspended solids**.

t-test—test of statistical significance based on the use of Student's t-distribution and used to compare two sample averages or a sample average and a hypothetical value.

TTO—abbreviation for total toxic organics.

TTT diagram—abbreviation for the time-temperature-transformation diagram used to describe the isothermal phase transformation of an alloy.

tubercles—knoblike protrusions of corrosion products.

tuberculation—corrosion process that produces localized corrosion products on the metal surface as knoblike prominences.

turbidity—reduction of transparency due to the presence of suspended particulates.

turbining—an abrasive cleaning method for tubes and pipelines that uses air, steam, or water to drive a motor that turns cutters, brushes, or knockers to remove deposits.

turbulence—irregular motion of a flowing fluid.

turbulent flow—fluid flow in which the velocity at a given point changes constantly in magnitude and its direction.

turning—generating external surfaces of revolution by the action of a cutting tool on a rotating workpiece, usually in a lathe.

turpentine—oily liquid extracted from pine resin and consisting primarily of a number of terpene hydrocarbons of the general formula $C_{10}H_{16}$. Used primarily as a solvent.

TWA—*abbreviation for* **Time-Weighted Average**.

two-metal corrosion—*see* **galvanic corrosion**.

Tyndall effect—scattering of light as it passes through a medium containing small particles.

U

UF—*abbreviation for* **ultrafiltration**.

ULC—abbreviation for Underwriters Laboratory (UL) Classification.

ultimate analysis—analytical determination of the proportions (amounts of and ratio) in which the elements are present in an organic substance

ultimate strength—maximum stress that a material can sustain. Also the tensile strength.

ultracentrifuge—apparatus for applying centrifugal forces up to a million times gravity causing very fine particles to settle out.

ultrafiltration (UF)—filtration technique whereby species are passed or rejected at a membrane surface on the basis of their size and that utilizes pressure as the driving force for the separation. The method can be used to separate large dissolved molecules (macromolecules) or colloids from a host substance.

ultrapure water—inexact term meaning water that has been processed beyond the usual pretreatment steps of settling, chlorination, filtration, and softening. Ultrapure water is water that has typically been treated by one or more of: distillation, deionization, reverse osmosis, sanitization (ozonation, ultraviolet radiation), or ultrafiltration. Ultrapure waters are commonly utilized in the preparation of pharmaceuticals, washing of electronic components during manufacture, nuclear power plant makeup waters, boiler feedwater, and the sterile water used for solution of drugs for injection in humans and other animals.

ultrasonic—mechanical vibrations of frequency greater than about 20,000 Hz.

ultrasonic cavitation test device—vibratory cavitation test device whose driving frequency is in the ultrasonic range (greater than 20,000 Hz). *See also* **hertz**.

ultrasonic cleaning—cleaning of surfaces by immersion in water, solvent cleaners, acid cleaners, or aqueous detergents, and applying ultrasonic vibrations to produce a cavitation effect with an attendant violent scouring action.

ultrasonic testing—ultrasonic technique used in quality control to measure thickness and also measure and locate flaws in metal caused by corrosion or erosion. Measurements can be made nondestructively from the outside of vessels and pipelines even during equipment operation.

ultraviolet absorber (UV absorber)—compound that absorbs ultraviolet radiation. *Also called* **ultraviolet stabilizer**.

ultraviolet radiation (UV)—invisible region of the electromagnetic spectrum from 10 to 380 nanometers.

ultraviolet regions—CIE division of the ultraviolet spectrum into three regions: UV-A, radiation in wavelengths between 315 and 400 nanometers (nm); UV-B, radiation in wavelengths between 280 and 315 nm; and UV-C, radiation in wavelengths shorter than 280 nm. *See also* **CIE**.

ultraviolet stabilizer (UV stabilizer)—additive to plastics that helps the plastic resist light degradation from exposure to ultraviolet radiation. Benzophenones and benzotriazoles are commonly used UV stabilizers. *Also called* **ultraviolet absorber.**

ultraviolet-visible spectroscopy (UV-visible spectroscopy)—technique for chemical analysis and the determination of structure utilizing the ultraviolet and visible regions of the electromagnetic spectrum to induce electronic transitions in molecules.

unconfined aquifer—aquifer occurring near the earth's surface with an impermeable layer of clay or rock as its lower boundary.

unctuous—having a greasy, oily, or soapy feeling when rubbed or touched by the fingers.

undercoat—any paint applied under a topcoat and over the primer or filler. The undercoat is highly pigmented for good hiding power and to give thickness to the paint film. Undercoat is sometimes also used as a synonym for primer.

undercoating—bituminous coating sprayed on the underside of automobiles to minimize rusting.

undercooling—unstable state in which a liquid remains liquid below its normal freezing point.

undercutting—gradual penetration and spread of corrosion beneath a coating from a break or pinhole in the film or from unprotected edges.

underfilm corrosion—corrosion occurring under paint coatings and other organic films at exposed edges or due to filiform corrosion.

undersaturated solution—solution that contains less of a solute than needed to saturate it.

underwater hull—portion of the hull of a ship that is normally fully immersed in the sea.

Unified Numbering System (UNS)—alphanumeric designation designed to identify any metal or alloy. It consists of a single uppercase letter followed by five digits.

uniform corrosion—corrosion that proceeds at about the same rate over the surface of a material. *Also known as* **general corrosion.**

unit cell—group of particles (atoms, ions, or molecules) in a crystal that is repeated in three dimensions in the crystal lattice.

universal indicator—mixture of acid-base indicators that changes color over a range of pH.

UNS—*abbreviation for* **Unified Numbering System**.

unsaturated—solvent capable of dissolving more of a solute at a given temperature. Also of or pertaining to a compound, especially of carbon, containing atoms that share more than one valence bond.

unsaturated zone—region between the ground surface and the water table that contains a mixture of air and water in its pore spaces.

upper atmosphere—upper part of the earth's atmosphere above about 300 kilometers.

urea plastics—plastics based on resins made from the condensation of urea and aldehydes.

urethane coatings—coatings containing a minimum of 10% by weight of a polyisocyanate monomer reacted to yield polymers containing urethane linkages, polyisocyanate groups, or active isocyanate groups. Classified in ASTM D 16-75 as one of six types according to their curing mechanisms: Type I One Package Prereacted Type; Type II One Package Moisture-Cured Type; Type III One Package Heat-Cured Type; Type IV Two Package Catalyzed Type; Type V Two Package Polyisocyanate Polyol-Cured Type; and Type VI One Package Nonreactive Lacquer Type.

urethane resins—synthetic resins containing the repeating group –NHCOO– and made by copolymerizing isocyanate esters with polyhydric alcohols. *See also* **polyurethanes**.

urine—aqueous fluid secreted from the blood by the kidneys, stored in the bladder, and discharged by the urethra. Normal adult urine mainly contains water, urea, creatinine, ammonia, uric acid, and small amounts of various inorganic salts.

USASI—abbreviation for the United States of America Standards Institute.

USDA—abbreviation for the U.S. Department of Agriculture.

useful life—length of time a coating is expected to remain viable in service. Also called service life.

USGS—abbreviation for the United States Geological Survey.

UV—*abbreviation for* **ultraviolet radiation**.

V

vacuum—space in which there is no gas or a very low pressure of gas, i.e., relatively few atoms or molecules. A perfect vacuum would contain no atoms or molecules.

vacuum coating—process for depositing metals and metal compounds from a source in a high-vacuum environment onto a substrate.

vacuum corrosion—term usually restricted to high-temperature deterioration of a material in a vacuum as a result of evaporation or volatilization. This form of deterioration is not properly described as corrosion.

vacuum deposition—condensation of thin metal coatings on a cool surface in a vacuum.

vacuum metallizing—process in which a metal or metal compound is evaporated at high temperature in a closed, evacuated chamber and then allowed to condense on a workpiece within the chamber.

valence—combining power of an atom or radical equal to the number of hydrogen atoms that the atom or radical could combine with or displace in a chemical compound. Hydrogen has a valence of 1. The valence is equal to the ionic charge in ionic compounds. In covalent compounds, it is equal to the number of bonds formed. *See also* **oxidation-reduction**.

valence electrons—electrons in the outermost energy level of an atom and that are gained, lost, or shared in chemical reactions.

Van der Waals force—attractive force between atoms or molecules. The force is much weaker than those arising from valence bonds and is inversely proportional to the seventh power of the distance between the atoms or molecules. It is the force largely responsible for the nonideal behavior of gases and for the lattice energy of molecular crystals.

vapor—gaseous phase of matter that normally exists in a liquid or solid state. Also applies to gas close to the liquid state, e.g., water vapor in the atmosphere. The terms vapor and gas are often used interchangeably.

vapor barrier—moisture-impervious layer or coating that prevents the passage of moisture or vapor into a material or structure.

vapor corrosion—acceleration of atmospheric corrosion by volatile contaminants such as short-chain organic acids, hydrogen chloride, sulfur dioxide, hydrogen sulfide, phenol, nitric acid, and ammonia, among others.

vapor degreasing—use of a condensing solvent vapor to dissolve and rinse away oil and grease from a metal surface prior to subsequent use or processing of the metal.

vapor deposition—*see* **chemical vapor deposition** and **physical vapor deposition**.

vaporization—conversion from liquid (or solid) to gaseous state. Direct conversion of solid to gaseous state bypassing the liquid state is known as sublimation.

vapor phase corrosion inhibitors (VPI)—solid or liquid volatile organic compounds having significantly high vapor pressure at room temperature and possessive of corrosion inhibitive behavior that are used within sealed enclosures to protect metallic articles. The most common VPI are dicyclohexylamine nitrite and cyclohexylamine carbonate. *Also known as* **volatile corrosion inhibitors (VCI)**.

vapor plating—process for the deposition of a metal or compound on a heated surface by reduction or decomposition of a volatile compound of the metal at a temperature below the melting point of the deposit and the base material.

vapor pressure—pressure exerted by the vapor of a solid or liquid when in equilibrium with the solid or liquid. Also that part of the atmospheric pressure created by water vapor.

variance—the mean of the squares of the variations from the mean of a frequency distribution. *See also under* **statistical terms**.

varnish—liquid composition that is converted to a transparent or translucent solid film after its application as a thin film. Also describes a deposit resulting from the oxidation and polymerization of fuels and lubricants that is similar to but softer than a lacquer deposit produced by similar processes.

Vaseline—trademark for refined petroleum jelly or petrolatum.

VCI—*abbreviation for* **volatile corrosion inhibitor**.

vegetable fats and oils—organic substances obtained by crushing and rendering the kernels, seeds, and fruit of specific plants. The oils are liquid at room temperature and contain varying percentages of unsaturated fat. The oils are classified as either drying or nondrying with the former forming tough, elastic films when exposed to the atmosphere due to oxygen absorption. Nondrying oils contain very little unsaturated fats and do not form such films.

vehicle—liquid portion of paint in which the pigment is dispersed. It is composed of binder and solvent or thinner. Anything else also dissolved in the paint is a part of the vehicle.

vehicle service test (fleet test)—ultimate test for evaluating engine coolant performance in which vehicles utilizing the coolant are operated under prevailing conditions. Also called fleet test.

velometer—instrument for measuring air velocity.

verdigris—green patina of basic copper salts formed on copper due to atmospheric corrosion. The basic copper salts are primarily basic carbonate ($CuCO_3 \cdot Cu(OH)_2$), basic sulfate ($CuSO_4 \cdot Cu(OH)_2 \cdot H_2O$), and, in some cases, the basic chloride ($CuCl_2 \cdot Cu(OH)_2$).

vermiculite—clay mineral of platelet form; chiefly, a hydrous silicate of magnesium, iron, and aluminum.

very thorough blast cleaning—*see* **white blasting**.

vibratory cavitation—cavitation caused by pressure changes within a liquid induced by the vibration of a solid surface immersed in the liquid.

vibratory cavitation erosion test—test method for determining the rate of cavitation erosion produced on different materials based on generating cavitation bubbles that collapse on the face of a test specimen vibrating at high frequency in the test liquid under specified test conditions. See ASTM G 32 for details.

vibratory finishing—process for deburring and surface finishing in which the product and an abrasive mixture are placed in a container and vibrated.

vinyl—unsaturated, univalent radical $CH_2=CH-$ derived from ethylene. Also any of the polyvinyl chloride or polyvinyl acetate family of coatings.

vinyl acetate plastics—plastics based on polymers of vinyl acetate or copolymers of vinyl acetate with other monomers.

vinyl butyral wash primer—*see* **wash primer**.

vinyl chloride plastics—plastics based on polymers of vinyl chloride or copolymers of vinyl chloride with other monomers.

vinylidene chloride plastics—plastics based on resins made by the polymerization of vinylidene chloride or copolymerization of vinylidene chloride with other unsaturated compounds.

vinyl plastics—plastics based on resins made from monomers containing the vinyl group.

viruses—particles of size below the resolution of the light microscope that are not cellular in structure and are composed mainly of nucleic acid surrounded by a protein sheath.

They lack metabolic machinery and can exist only as intracellular, highly host-specific parasites.

viscometer—instrument for measuring the viscosity of a fluid. The chief types of viscometers are classified as: capillary, rotational, outflow or orifice type, falling ball, and radial flow.

viscosity—resistance to flow exhibited within the body of a fluid when it is subjected to shear stress. *Opposite of* **fluidity**.

visible light—electromagnetic radiation in the 400 to 700 nanometer wavelength range.

vitreous—having a glasslike appearance.

vitreous enamel—essentially glass coating compositions of suitable coefficient of expansion that are fused mainly on steel, but also on copper, brass, and aluminum. *Also called* **porcelain enamel**.

vitreous silica (fused quartz)—high-temperature corrosion-resistant material used for furnace muffles, burners, piping, and reaction chambers.

vitreous state—solid condition of matter resulting from the rapid cooling of a liquid such that crystals do not form at a definite temperature, but the viscosity increases steadily until a glassy (disordered amorphous solid) substance is obtained.

vitrification—progressive reduction in porosity of a ceramic composition as a result of heat treatment.

VM&P naphtha—abbreviation for varnish maker's and painter's naphtha. A low solvent power, flammable hydrocarbon solvent, consisting primarily of aliphatic hydrocarbons, obtained as narrow boiling-range fractions of petroleum having a boiling range of about 93 to 149 °C (200 to 300 °F). Also called painter's naphtha and benzine.

VOC—*abbreviation for* **volatile organic compound**.

void—cavity unintentionally formed in a cellular material and substantially larger than the characteristic individual cells. In paint coatings, a term describing a holiday, hole, or skip in the film.

volatile—describes a liquid that evaporates readily at relatively low temperatures.

volatile content—percentage of materials in a paint that evaporate on drying or curing of the film.

volatile corrosion inhibitor (VCI)—*see* **vapor phase corrosion inhibitors**.

volatile matter—products, exclusive of moisture, given off by a material as gas or vapor using a prescribed test method.

volatile oil—oil distinguished from a fixed oil by the fact that a drop of the former does not leave a spot on paper on relatively standing. Volatile oils are principally plant oils.

volatile organic compound (VOC)—any organic compound that participates in atmospheric photochemical reactions as designated by EPA standards. A group of chemicals that react in the atmosphere with nitrogen oxides in the presence of heat and sunlight to form ozone.

volatile solvent—any nonaqueous liquid that evaporates readily at room temperature and atmospheric pressure.

volt—symbol V. The SI unit of electric potential, potential difference, or emf and defined as the difference of potential between two points on a conductor carrying a constant current of one ampere when the power dissipated between the points is one watt.

voltaic cell—*see* **galvanic cell**.

voltaic pile—battery devised by Volta consisting of a number of flat voltaic cells joined in series where the liquid electrolyte is absorbed into paper or leather disks.

voltmeter—instrument used to measure electrical current in volts (voltage).

volume—space occupied by a body or mass of fluid.

volume resistivity—resistance in ohms per unit of volume. The property of a material that determines its resistance to the flow of an electrical current.

volumetric analysis (titrimetric analysis)—quantitative analytical method wherein solutions of known concentration are reacted in some fashion with the sample to determine the concentration of the unknown component of the sample.

VPI—*abbreviation for* **vapor phase corrosion inhibitors**.

vulcanization—irreversible process by which a rubber compound becomes less plastic and more resistant to swelling by organic liquids while its elastic properties are improved or extended over a greater range of temperature and that results from change in its chemical structure. Vulcanization is achieved by heating the rubber with sulfur or sulfur-containing compounds.

W

wallboard—monolithic board material used in interior construction and made from such as pressed cellulose fibers, preformed plaster, cement-asbestos, and plywood. Used in place of plastered interior surfaces.

walnut shells—one of several agricultural products used as a mild abrasive in abrasive blast cleaning. Others include crushed pecan shells, rice hulls, rye husks, corncobs, and sawdust. Smooth plastic beads serve a similar purpose.

warm lime softening—precipitation softening process for water accomplished at temperatures in the range of 49 to 60 °C (120 to 140 °F). Precipitation reaction rates are increased compared with cold lime softening. *See also* **cold lime softening**.

Warren test—timed immersion test in a mixture of 10% nitric acid and 3% hydrofluoric acid at 70 to 80 °C (158 to 176 °F) to evaluate type 316L stainless steel for excessive precipitated chromium carbide content. The immersion severely attacks material with precipitated chromium carbide but not with sigma phase chromium.

wash coat—very thin, semitransparent coat of paint applied as a preliminary coating on a surface to act as a sealer or guide coat.

washing—cleaning in an aqueous medium. Also the erosion of a paint film after rapid chalking.

washing soda—sodium carbonate decahydrate ($Na_2CO_3 \cdot 10H_2O$).

wash primer—two-component corrosion-inhibiting primer paint based on an inhibiting chromate pigment and phosphoric acid in a polyvinyl butyrate binder system. It is used as the principal pretreatment or metal conditioner of the maintenance painting industry and commonly designated as the WP-1 Wash Primer. It consists of approximately 9% polyvinyl butyral and 9% zinc tetroxychromate in a mixture of isopropanol and butanol as one solution, which is mixed just before using with a solution of 18% phosphoric acid in isopropanol and water in the weight ratio of four of the pigmented solution to one of the latter. Single-component wash-primer systems have also been developed utilizing less soluble inhibitors such as lead chromate and chromic phosphate, but they are not nearly so widely used as the two-component system. Wash primers are used on steel, aluminum, zinc, tin, stainless steel, and titanium. Also called etch primer.

wastage—cooling tower term describing the total "concentrated" water that is removed from the system and replaced by fresh makeup water. *See also* **blowdown** and **cycles of concentration**.

waste—material that is removed, rejected, or otherwise lost in manufacturing.

waste ash—fly ash and bottom ash generated from using coal as fuel in fossil fuel power plants.

waste-heat boilers—steam-generating boilers utilizing the exotherms from chemical or plant processes for purposes of economy and energy conservation.

wastewater—water discharged from a process as a result of its formation or use in that process.

wastewater-treatment plants—municipal treatment facilities that generally consist of: (1) primary treatment where holding tanks or ponds are used to allow sewage to settle and solids are removed as sludge, (2) secondary treatment where the effluent from (1) passes through sand and gravel filters, and bacteria convert most of any residual organic matter to stable inorganic materials, and where chlorine is usually added to kill any remaining pathogenic microorganisms, and finally (3) advanced or tertiary treatment may follow where charcoal filtration may be used to absorb any organic molecules not previously removed.

water absorption—ratio of the weight of water absorbed by a material to the weight of the dry material.

water-base hydraulic fluids—any of three basic types of hydraulic fluids; water-glycols with about 40% water, water-in-oil emulsions (invert emulsions) with about 40% water, and high water base fluids (95-5 fluids) with about 95% water. The latter are further subclassified as either solutions (chemical or synthetic solutions), emulsions (soluble oils), or microemulsions.

water-base mud—common, conventional oil- or gas-well drilling fluid where water is the suspending medium for mud solids and is the continuous phase, whether or not oil is present. *See also* **oil-base mud**.

water blasting—blast cleaning of metal using high-velocity water.

waterborne coatings—latex paints and paints containing water soluble binders. Coatings that bear or carry water and are water reducible. *Also called* water-based coatings and **water-reducible coatings**. *Compare with* **solvent-borne coatings**.

waterborne foulants—suspended materials such as mud, sand, silt, clay, biological matter, and oil contained in some waters and that may cause fouling.

water-break—discontinuous film of water on a surface signifying nonuniform wetting and usually associated with a surface contamination. *See also* **cleanliness test**.

water conditioning—broad term covering all processes used in treating water to remove or reduce undesirable impurities to specified tolerances. *See also* **water impurities**.

water contaminants—contaminants whose introduction into water supplies are related to rainfall, the geologic nature of the watershed or underground aquifer, and the activities of nature and the human population. The contaminants include both dissolved matter and insoluble constituents. The dissolved matter includes solubilized minerals and organic matter and also the transient constituents affecting water acidity-alkalinity, cyclical biological constituents, and oxidizing-reducing materials from the natural environment and from human treatment of water. Insoluble constituents include floating, suspended, and settling solid matter of mineral and organic basis and microbial organisms such as algae, bacteria, fungi, and viruses.

water-dispersible coating—organic coatings that normally are solvent-based, but by adjusting the chemistry can be dispersed in water.

water-displacing rust preventives—compositions that displace water from surfaces and used primarily in the rust protection of mechanical equipment in daily operation at machine shops and factories and in the reconditioning of water-wet apparatus such as electrical equipment.

water-drive reservoirs—oil-field recovery process. Where oil fields exist in formations associated with large quantities of salt water, which is under pressure, the energy associated with the salt water assists in producing the oil. *See also* **dissolved-gas-drive oil reservoirs, gas-cap-drive oil reservoirs**, and **oil-well pumps**.

waterflooding—secondary oil recovery process. A process of displacing oil from underground formations with injected water to force its movement to the surface for recovery.

water-formed deposit—any accumulation of insoluble materials from a cooling water including scale, sludge, foulants, sediments, corrosion products, or biological deposits.

water gas—mixture of carbon monoxide and hydrogen produced by passing steam over hot carbon or coke. Used as a fuel gas.

water glass—glassy sodium or potassium silicate that is soluble in water. Also a viscous colloidal solution of sodium or potassium silicates in water.

water hardness—water classification by its calcium ion content as being either soft or hard. Generally, soft water contains Ca^{2+} ions at 60 ppm or less calculated as $CaCO_3$, while hard water contains 60 ppm or more. Metals generally corrode faster in soft than in hard water, because CO_3^{2-} ions form a protective $CaCO_3$ layer on metallic surfaces in hard water. *See also* **hard water**.

water immersion—exposure condition in which the material is in direct contact with fresh or salt water.

water impurities—undesirable constituents found in water. They may be grouped under three headings: (1) suspended matter, color, and organic matter including sediment, turbidity, color, microorganisms, tastes, odors, and other organic matter; (2) dissolved mineral matter consisting chiefly of the bicarbonates, sulfates, and chlorides of calcium, magnesium, and sodium; and (3) dissolved gases. Other common impurities, found either as suspended or dissolved matter, include silica, alumina, iron, manganese, fluorides, nitrates, potassium, and sulfuric acid.

waterline—area of the hull of a ship immediately above and below the water surface that can be alternately exposed to the air or immersed in water depending on the weather and the loading of the ship.

waterline attack—corrosive attack of metals partially immersed in an aqueous system with air above. The metal undergoes simultaneous differential aeration cell and crevice corrosion at the waterline junction.

water of crystallization—water present in crystalline compounds in definite proportions.

water, natural—water as it occurs in nature.

water paint—paint whose vehicle is a water emulsion, water dispersion, or utilizes ingredients that react chemically with water.

waterproofing—treatment to prevent passage of water.

water, pure—in laboratory terms, pure water is defined as reagent grade water. Within this general definition, however, reagent grade water may have different levels of purity as described by its specific conductance, specific resistance, total matter, silicate content, potassium permanganate reduction potential, culture colony count, and pH. ASTM specifies three levels of reagent grade water purity: Types I, II, and III with Type I the most pure. See ASTM D 1193 for the standard specification for these types.

water-reducible coatings—water soluble types or lattices or emulsions of coating systems, all of which can be reduced (diluted) with water, water-cosolvent mixtures, and, sometimes, with alkali in the case of alkali-soluble resins. *Also called* **waterborne coatings**.

water-repellent—treatment to provide resistance to penetration by water.

water-resistance—ability to retard penetration or wetting by water.

waterside deposit—hard encrustation, usually rich in sulfate or carbonate salts of calcium, deposited on the surfaces of boilers and equipment through which cooling water is passed. In addition to these, deposits can result from treatment chemicals (dispersants, sludge conditioners, corrosion inhibitors), silica, and process chemical leaks.

water softening—to remove hardness from water.

water soluble oils (emulsions, water emulsifiable oils, emulsifiable cutting fluids)—metalworking fluids that are emulsified suspensions of oil droplets in water. They may also contain organic sulfur, chlorine, or phosphorus materials to improve their lubricating ability.

water table—upper surface of an unconfined aquifer at atmospheric pressure.

water table aquifer—*see* **unconfined aquifer**.

water types—water is usually described in terms of its nature, usage, or origin. Ground waters originate in subterranean locations while surface waters comprise the lakes, rivers, and seas. Fresh water is either from surface or ground sources, and typically contains less than 1% dissolved salt. Brackish water contains between 1 and 2.5% sodium chloride. Seawater typically contains about 3.4% sodium chloride and has a slightly alkaline pH 8. Steam condensate describes the water resulting from industrial steam condensation. Boiler feedwater makeup is always presoftened and deaerated and may include some steam condensate. Potable water is fresh water that is sanitized to make it safe for drinking and culinary purposes.

water vapor transmission (WVT)—rate of water vapor flow, under equilibrium conditions, through a unit area of a material between its parallel surfaces and normal to the surface.

water white—descriptive term for transparent liquids and solids that approach the colorless nature of water.

watt—unit of active power. One watt is energy, work, or quantity of heat expended at a rate of one joule per second. Symbol is W.

wavelength—repetition length of a wave. Defined as the *x*-distance between any two successive corresponding points on the wave.

wax—any substance physically resembling beeswax and that is unctuous, insoluble in water, but soluble in most organic solvents. The term is generally applied to all waxlike solids and to certain liquids composed of monohydric alcohol esters. Waxes come from various animal, vegetable, mineral, or synthetic sources. They differ from fats in being less greasy, harder, polishable. Mineral waxes are mixtures of hydrocarbons with high molecular weights. Animal and vegetable waxes are mainly esters of fatty acids. The distinction between an industrial oil and wax is a matter of viscosity and not chemical composition.

weak acid—acid only partially dissociated in aqueous solution. *Opposite of* **strong acid**.

weak base—base that dissociates or ionizes to a small degree; base which produces few OH⁻ ions in solution. *Opposite of* **strong base**.

wear—damage to a solid surface involving progressive loss of material due to relative motion between that surface and a contacting surface or substance. Also the removal of material or impairment of surface finish through friction or impact. It is an artificial process and its rate may be affected by chemical action. *See also* **erosion** and **abrasion**.

wear corrosion   simultaneous action of wear and corrosion that accelerates material loss due to their synergistic effect. *See also* **fretting corrosion**.

weather—state of atmospheric conditions; includes conditions of humidity, precipitation, temperature, cloud cover, visibility, pressure, and wind at any one place and time. Also to age, deteriorate, discolor, etc., as a result of exposure to the weather.

weather deck—deck and walking surface of a ship that are fully exposed to the weather.

weathering—process of disintegration and decomposition resulting from exposure to the atmosphere, its chemical action, sunlight, frost, water, and heat. Also method of removing mill scale from heavy structural steel members by exposing them outdoors for a period of some months to allow mill scale to crack off under the stress of expansion and contraction.

weathering steels—low-alloy steels containing small amounts of copper (tenths of one percent) or nickel and chromium and that have increased resistance to atmospheric corrosion because they form a tighter, more protective rust film. They are primarily used for buildings, bridges, and guard rails.

weatherometer—testing apparatus in which specimen materials can be subjected to artificial and accelerated weathering intended to simulate natural atmospheric weathering but in shorter time.

wedging action—stress generated by corrosion products in constricted regions, e.g., at a crack tip. *See also* **denting**.

weep hole—deliberately placed small hole in a wall or assembled structure allowing for drainage of water.

weight—force by which a body is attracted to the earth. *See also* **mass**.

weight-loss method—direct corrosion monitoring method that measures the rate of corrosion by metallic weight loss.

weld—metallic bond between like or unlike metals joined by fusion.

weld decay—intergranular corrosion, usually of stainless steels and nickel alloys, because of metallurgical changes in the heat-affected zone during welding.

weld slag—amorphous deposits formed during welding.

Wenner method (four-pin method)—method of measuring soil resistivity wherein four metal rods are driven into the ground equally spaced along a straight line, then current is caused to flow between the outer pair. The potential between the inner pair is measured, and soil resistivity is calculated.

Weston cell (cadmium cell)—primary voltaic cell consisting of a mercury anode covered with a paste of cadmium sulfate and mercurous sulfate, and a cathode of cadmium mercury amalgam covered with cadmium sulfate, and utilizing a saturated solution of cadmium sulfate as the electrolyte. Used as a standard cell producing an emf of 1.0186 volts at 20 °C (68 °F).

wet-and-dry-bulb hygrometer—*see* **hygrometer**.

wet blasting—process for cleaning or finishing by means of a slurry of abrasive in water that is directed at high velocity against the workpiece.

wet-bulb temperature—dynamic equilibrium temperature attained by a water surface when the rate of heat transfer to the surface by convection equals the rate of mass transfer away from the surface. It is the temperature indicated by a wet-bulb psychrometer and is used in determining relative humidity and dew point.

wet chemical analysis—classical laboratory manual methods, conducted on a macro- or microchemical scale, used to separate and identify (qualitatively and/or quantitatively) the components of matter. The equipment utilized, essentially comprised of analytical balances and laboratory glassware, tends to be of a universal nature. The methods do not generally employ any analytical instrumentation. Wet chemical analysis includes gravimetric analysis and volumetric analysis methods for the quantitative analysis of matter.

wet corrosion—corrosion occurring when a liquid is present. Usually involves aqueous solutions or electrolytes and is the most common form of corrosion. *Contrast with* **gaseous corrosion**.

wet/dry tower—cooling tower design where the supersaturated air is either heated going through finned tube heat exchangers in a series path design or it is diluted in a large volume of warmer, drier air in a parallel path design. Used for plume abatement.

wet film thickness—thickness of a liquid film, such as paint, immediately after its application.

wet milling—grinding of materials with abrasive and sufficient liquid to form a slurry.

wet rot—decay of timber caused by fungi that flourish in alternate wet and dry conditions.

wet scrubbers—industrial systems used to remove solid, liquid, or gaseous contaminants from a gas stream to control air polluting emissions.

wet sponge tester—low voltage holiday detection device.

wet steam—steam that contains water droplets.

wet storage stain—white corrosion products on newly galvanized, bright surfaces, especially in the crevices between closely packed sheets and angle bars if the surfaces come into contact with condensate or rainwater and drying is retarded. *Also called* **white rust**.

wetting—adhesion of a liquid to the surface of a solid. Also the ability of a liquid to spread uniformly and rapidly over the surface of a material.

wetting agent—substance that reduces the surface tension of a liquid causing it to spread more readily on solids. *See also* **surface-active agent**.

Wheatstone bridge—electrical circuit for measuring the value of a resistance.

whiskers—slowly developing, very fine, whiskerlike metal growth on relatively pure low-melting-point tin, cadmium, and zinc coatings that are under stress, usually internal, resulting from plating either at low temperatures, or from solutions containing brightening agents. Their growth relieves stress. The major problem associated with whiskers is electrical shorting across faces of capacitor plates and between conductors in microswitches and microcircuits.

white blasting—blast cleaning of steel surface such that all contamination, scale, and rust are removed and the surface exhibits the gray and white luster of virgin steel. This surface preparation is now designated blast cleaning to visually clean steel by the ISO. Near-white blasting results in a surface exhibiting less than 5% of the contaminate residues. This surface preparation is now termed very thorough blast cleaning by the ISO. Commercial blasting is most generally used in maintenance painting and refers to a surface free of all contaminants and two-thirds of its unit surface is free of all visible residues. Slight discolorations and shadows remaining on the steel after the removal of adherent rust and mill scale, are allowed over the other one-third of the surface. This surface preparation is now termed thorough blast cleaning by the ISO. Brush blasting is usually a field method of cleaning. It results in a surface free of oil, grease, dirt, loose rust scale, loose mill scale, and loose paint, but tightly adhering mill scale, rust, paint, and other coating remains. This surface preparation is now termed light blast cleaning by the ISO.

white frost—*see* **hoar frost**.

white iron (white cast iron)—extremely hard and brittle cast iron with substantially all its carbon in solution and in combined form.

white liquor—cooking liquor from the kraft pulping process produced by recausticizing green liquor with lime.

white metal blast—*see* **white blasting**.

white oils—light-colored and usually highly refined mineral oils used for lubrication.

white rot—internal attack of construction wood used in cooling towers resulting from deterioration of its lignin binder and caused by thermophilic fungi such as the *Basidiomycetes* group.

white rust—white corrosion products, e.g., zinc hydroxide, zinc oxide, zinc carbonate, formed on zinc-coated articles. *See also* **wet storage stain**.

white spirit—British term for a liquid mixture of hydrocarbons obtained from petroleum and used primarily as a solvent for paint. Called mineral spirits in the United States. A turpentine substitute.

white wash—paint composition based on lime or whiting bound with glue, casein, or water-dispersible binders.

white water—water that drains through the endless wire belt of the fourdrinier paper machine and collects in the wire pit. It contains soluble chemicals used in the papermaking process, soluble by-products, and escaped pulp fibers.

whiting—calcium carbonate ($CaCO_3$) of higher purity than the natural product.

WHO—abbreviation for the World Health Organization.

wicking—absorption of a liquid by capillary action.

wind—motion of air relative to the earth's surface.

windage—*see* **drift**.

windage loss—water droplets from a cooling tower that are carried into the atmosphere due to drawing air through the tower. This water loss is part of the total blowdown loss of the cooling tower system. *Also known as* **drift loss**.

wiping—wear term describing the smearing or removal of material from one point, and often redeposition, on the surface of two bodies in sliding contact.

wire brush—hand cleaning tool comprised of bundles of wire.

wire-on-bolt test—*see* **CLIMAT test**.

wood—hard structural and water-conducting tissue that is found in many perennial plants and forms the bulk of trees and shrubs. It is composed of xylem and associated cells, such as fibers.

wood alcohol—methanol (CH_3OH), derived from the destructive distillation of wood.

wood deterioration—refers to the chemical (delignification), biological (surface rot and internal decay), and physical (erosion and thermal) deterioration wood used in cooling-tower construction may experience.

wood preservative—material applied to wood to prevent its destruction by fungi, wood-boring insects, marine borers, or fire. The preservatives principally consist of preservative oils, toxic chemicals in organic solvents, and water-soluble salts.

wood preservative coating—coating formulated to protect wood from decay and insect attack.

work—product of a force and the distance moved by its point of application acting on a body.

Worker Right-to-Know Law—*see* **Hazard Communication Standard**.

work hardening—increase in the hardness of a metal as a result of working it cold. Strain hardening.

working electrode—specimen electrode in an electrochemical cell.

working electrode potential—measured potential of a specimen electrode when its interface is crossed by a net current.

workover fluid—any fluid used in the workover operation of a well.

wrinkling—formation of a surface appearance in a coating film that resembles a prune skin.

wrought iron—highly refined form of iron containing less than 0.03% carbon and with 1 to 3% of slag (mostly as iron silicate) that is evenly distributed throughout the material in threads and fibers so that the product has a fibrous structure quite dissimilar to that of crystalline cast iron. It is a "mechanical" mixture of slag and low-carbon steel. Produced by burning out the carbon and other impurities from pig iron. It generally rusts less readily than other forms of metallic iron, and welds and works more easily.

wustite—FeO component of scale on steel that can form up to 90% of the scale. It dissolves readily in acid pickling. *See also* **magnetite** and **scale**.

WVTR—abbreviation for water vapor transmission rate; a value usually expressed in $g^{-2} d^{-1}$ per day for barrier wrap materials. *See also* **barrier materials classification** and **water vapor transmission**.

X Y Z

XPS—*abbreviation for* **X-ray photoelectron spectroscopy**.

X-radiation (x-radiation)—electromagnetic radiation having a wavelength about $\frac{1}{1000}$th that of visible light, i.e., shorter than ultraviolet but longer than gamma rays.

X-ray crystallography—use of X-ray diffraction to determine the structure of crystals or molecules.

X-ray diffraction—diffraction of X-rays by a crystal.

X-ray fluorescence—emission of X-rays from excited atoms produced by the impact of high-energy electrons, other particles, or a primary beam of other X-rays.

X-ray photoelectron spectroscopy—analytical technique for study of the surfaces of materials. Also known as electron spectroscopy for chemical analysis (ESCA). XPS uses the photoelectric effect to study the electronic structure of surfaces as well as the chemical identity of surface components.

X-rays (x-rays)—electromagnetic radiation of shorter wavelength than ultraviolet radiation produced by bombardment of atoms by high-quantum energy particles. The range of X-ray wavelengths is 10^{-11} to 10^{-9} meters.

yeasts—term describing some fermentative species of unicellular sac fungi (*Ascomycetes*). Yeasts generally require more moisture for growth than molds.

yellow brass—copper zinc alloy containing about 30% zinc. The alloy gradually dezincifies in seawater and in soft fresh waters, but this can be retarded by addition of 1% tin with the resulting alloy being called admiralty metal.

yellowing—development of yellow color or cast in a white pigment or colorless binder due to aging.

yield point—point on the load-elongation or stress-strain curve at which load or stress ceases to be linearly proportional to the elongation or strain.

yield strength—stress at which a material exhibits a specified deviation from proportionality of stress and strain.

yield stress—maximum shear stress that can be applied without causing permanent deformation.

Young's modulus—for perfectly elastic materials, it is the ratio of change in stress to change in strain within the elastic limits of the material.

ZDDP—*abbreviation for* **zinc dialkyl dithiophosphate**.

zeolite—natural or synthetic hydrated aluminosilicate with an open three-dimensional crystal structure in which water molecules are held in cavities in the lattice. The water can be driven off by heating and the zeolite can then absorb other molecules of suitable size. Used for ion-exchange water softening and as a refinery cracking catalyst.

zero convention—convention by which a potential equal to 0 is arbitrarily set for the standard hydrogen electrode (SHE) for the reaction $2H^+ + 2e^- = H_2$. All other potentials then are expressed with respect to this reference potential and assigned their numbers on the scale of electrode potentials.

zero potential—potential that exists under reversible conditions between hydrogen gas at one atmosphere and hydrogen ions at unit activity. This standard hydrogen half-cell is the electrode against which other half-cell potentials are quoted and is the accepted standard zero potential. *See also* **zero convention**. The standard hydrogen electrode is also arbitrarily defined as at zero potential for all temperatures.

zero-resistance ammeter—potentiostat programmed for zero potential difference between the reference and test electrode, and in addition, the current lead to the counterelectrode is connected to the reference electrode.

zero water discharge—practice for the continuous reuse of process waters by separating dissolved solids from the water.

zeta potential—difference in voltage between the surface of the diffuse layer surrounding a colloidal particle and the bulk liquid beyond. This electric potential causes colloidal particles to repel each other and stay in suspension. The zeta potential is a measure of this force. *See also* **electrokinetic potential**.

zincate—salt formed in solution by dissolving zinc or zinc oxide (ZnO) in alkali. The ionic formula for zincate is written ZnO_2^{2-}.

zincating—immersion coating of aluminum-base materials with zinc to facilitate electroplating of other metals on the aluminum article. Zincating is a chemical replacement in which aluminum ions replace zinc ions in an aqueous solution of zinc salts, thus depositing a thin, adherent film of metallic zinc on the aluminum surface. Adhesion of the zinc depends on metallurgical bonding.

zinc chromate—anticorrosive paint pigment whose nominal composition approximates $4ZnO \cdot K_2O \cdot 4CrO_3 \cdot 3H_2O$. *Also called* **zinc yellow** in the United States and **basic zinc chromate** in Europe.

zinc coated (galvanized)—generic term covering barrel galvanized, dipped, electrogalvanized, electrozinc plated, flake-galvanized, hot dipped, hot dip galvanized, hot dipped zinc-coated, peen-coated, mechanically plated, peen-galvanized, tumbler-galvanized, and wean-galvanized iron or steel articles.

zinc dialkyl dithiophosphate (ZDPP)—substance widely used as an EP agent, antiwear agent, and oxidation inhibitor in lubricating oils.

zinc dust—finely divided zinc metal used as a pigment in protective paints for iron and steel.

zinc embrittlement—form of liquid-metal embrittlement of austenitic stainless steels most commonly seen in the welding or fire exposure of these steels while in contact with galvanized steel components.

zinc hot dipping—*see* **galvanizing**.

zincing—*see* **zinc shot blasting**.

zinc oxide—ZnO. White pigment used in paint primarily for imparting mildew resistance and film reinforcing properties.

zinc phosphate coating—treatment used on steel to improve subsequent coating adhesion. *See also* **conversion coating**. Also a thin inorganic deposit of zinc phosphate on zinc or zinc-coated surfaces resulting from treatment with phosphoric acid.

zinc phosphate pigment—corrosion-inhibitive paint pigment of the composition $Zn_3(PO_4)_2 \cdot 2H_2O$.

zinc-plated—thin, electrochemical deposit of zinc on a surface as a result of immersion in an electrolytic bath, or as a mechanical deposit from peening with zinc.

zinc-rich primer—primer for ferrous metals having zinc dust at a concentration sufficient to make the dried film electrically conductive in order to provide cathodic protection to the ferrous substrate.

Zincrometal—two-coat proprietary coating system applied on steel on a continuous coil-coating line and resulting in a mixed-oxide underlayer containing metallic zinc particles and a zinc-rich organic (epoxy) topcoat that is weldable, formable, paintable, and compatible with commonly used adhesives.

zinc shot blasting (zincing)—blast cleaning procedure in which metallic zinc particles are used for all or part of the shot, grit, or sand.

zinc silicate primers—inorganic zinc-rich primer paints that utilize a silicate binder.

zinc tetroxy chromate—*see* **basic zinc chromate**.

zinc whiskers—*see* **whiskers**.

zinc yellow—U.S. term for the anticorrosive zinc chromate pigment ($4ZnO \cdot K_2O \cdot 4CrO_3 \cdot 3H_2O$). In Europe, this pigment is called basic zinc chromate. *See also* **zinc chromate** and **basic zinc chromate**.

ZP-10—trademark for an anticorrosive pigment nominally composed of zinc phosphate.

ZPA—trademark for an anticorrosive pigment chemically described as basic aluminum zinc phosphate hydrate.

zwitterion (ampholyte ion)—ion having a positive and negative charge on the same group of atoms. The zwitterion is most common with organic compounds that contain both acidic and basic groups in their molecules.

zymolytic reaction—chemical reaction catalyzed by an enzyme, especially bond rupture or splitting, and usually a hydrolysis.